Lyman B Tefft

Curiosities of Heat

Lyman B Tefft
Curiosities of Heat
ISBN/EAN: 9783743394933
Manufactured in Europe, USA, Canada, Australia, Japa
Cover: Foto ©berggeist007 / pixelio.de

Manufactured and distributed by brebook publishing software (www.brebook.com)

Lyman B Tefft

Curiosities of Heat

CURIOSITIES

OF

HEAT.

BY

Rev. LYMAN B. TEFFT.

PHILADELPHIA
THE BIBLE AND PUBLICATION SOCIETY,
530 Arch Street.

CURIOSITIES OF HEAT.

BY
REV. LYMAN B. TEFFT.

PHILADELPHIA:
THE BIBLE AND PUBLICATION SOCIETY,
530 ARCH STREET.

Entered according to Act of Congress, in the year 1871, by
THE BIBLE AND PUBLICATION SOCIETY,
In the office of the Librarian of Congress, at Washington.

WESTCOTT & THOMSON,
Stereotypers, Philada.

CONTENTS.

CHAPTER I.
MR. WILTON'S BIBLE CLASS.................................... 7

CHAPTER II.
NEW THOUGHTS FOR THE SCHOLARS....................... 26

CHAPTER III.
A DIFFICULT QUESTION... 58

CHAPTER IV.
HEAT A GIFT OF GOD... 83

CHAPTER V.
CONVEYANCE AND VARIETIES OF HEAT..................... 100

CHAPTER VI.
MANAGEMENT AND SOURCES OF HEAT...................... 120

CONTENTS.

CHAPTER VII.
PRESERVATION AND DISTRIBUTION OF HEAT 152

CHAPTER VIII.
MODIFICATION OF TEMPERATURE 176

CHAPTER IX.
THE MINISTRY OF SUFFERING 190

CHAPTER X.
TRANSPORTATION OF HEAT 213

CHAPTER XI.
AN EFFECTIVE SERMON 233

CHAPTER XII.
TRANSFER OF HEAT IN SPACE 254

CHAPTER XIII.
OCEAN CURRENTS AND ICEBERGS 272

CHAPTER XIV.
COMBUSTION.—COAL-BEDS 292

CONTENTS.

CHAPTER XV.
ECONOMY OF HEAT... 305

CHAPTER XVI.
A DAY OF JOY AND GLADNESS................................. 320

CURIOSITIES OF HEAT.

CHAPTER I.

MR. WILTON'S BIBLE CLASS.

"THE book of Nature is my Bible. I agree with old Cicero: I count Nature the best guide, and follow her as if she were a god, and wish for no other."

These were the words of Mr. Hume, an infidel, spoken in the village store. It was Monday evening. By some strange freak, or led by a divine impulse, he had determined, the previous Sunday afternoon, to go to church and hear what the minister had to say. So the Christian people were all surprised to see Mr. Hume walk into their assembly—a thing which had not been seen before in a twelvemonth. Mr. Hume did not shun the church from a dislike of the minister. He believed Mr. Wilton to be a good man, and he knew him to be kind and earnest, well instructed in every kind of knowledge and mighty in the Scriptures. He kept aloof because he hated the Bible. He had been instructed in the Scriptures when a boy, and many Bible truths still clung to his memory which he would have been glad to banish. He could not forget those stirring words which have come down to us from the Lord Jesus, and from prophets and apostles, and they sorely troubled his conscience. He counted the Bible an enemy, and determined that he would not believe it.

At that time there was an increasing religious interest in the church. Mr. Wilton had seen many an eye grow tearful as he unfolded the love of Christ and urged upon his hearers the claims of the exalted Redeemer. He found an increasing readiness to listen when he talked with the young people of his congregation. The prayer-meetings were filling up, and becoming more interesting and solemn. The impenitent dropped in to these meetings more frequently than was their wont. Mr. Wilton himself felt the power of Christ coming upon him and girding him as if for some great spiritual conflict. His heart was filled with an unspeakable yearning to see sinners converted and Christ glorified. He seemed to himself to work without fatigue. His sermons came to him as if by inspiration of the Holy Spirit. He felt a new sense of his call from God to preach the gospel to men, and spoke as an ambassador of Christ, praying men tenderly, persuadingly, to be reconciled to God, yet as one that has a right to speak, and the authority to announce to man the conditions of salvation.

A few of the spiritual-minded saw this little cloud rising, but the people in general knew nothing of it. Least of all did Mr. Hume suspect such an undercurrent of religious interest; yet for some reason, he hardly knew what, he felt inclined to go to church.

That afternoon the preacher spoke as if his soul were awed, yet lifted to heavenly heights, by the presence of God and Christ. Reading as his text the words, "Thou thoughtest that I was altogether such an one as thyself" (Ps. l. 21), he showed, first, the false notions which men form of God, and then unfolded, with great power and pungency, the Scripture revelation of the one infinite, personal, living, holy, just, and gracious Jehovah. This was the very theme which Mr.

Hume wished most of all not to hear. That very name, Jehovah, of all the names applied to God, was most disagreeable; it suggested the idea of the living God who manifested himself in olden time and wrought wonders before the eyes of men. But the infidel, with his active mind, could not help listening, nor could he loosen his conscience from the grasp of the truth. Yet he could fight against it, and this he did, determined that he would not believe in such a God—a God who held him accountable, and would bring him into judgment in the last great day. In this state of mind he dropped into Deacon Gregory's store.

Deacon Gregory was accustomed to obey Paul's injunction to Timothy: "Be instant in season, out of season; reprove, rebuke, exhort with all long suffering and doctrine." Having taken Mr. Hume's orders for groceries, he said, "I was glad to see you at church yesterday, Mr. Hume. How were you interested in the sermon?"

"I like Mr. Wilton," answered Mr. Hume; "I think him a very earnest and good man."

"But were you not interested and pleased with the discourse? It seems to me that I shall never lose the impression of God's existence and character which that discourse made upon me. I almost felt that Mr. Wilton spoke from inspiration."

"I suppose he was inspired just as much as the writers of that book which men call 'the Bible.'"

"But can you wholly get rid of the conviction that the Bible is the word of God, written by holy men inspired by the Holy Spirit?"

"You know, Deacon Gregory, that I do not believe what you profess to believe. The book of Nature is my Bible. I agree with old Cicero: I count Nature the best guide, and follow her as if she were a god, and wish no other."

Deacon Gregory had never read Cicero, and of course did not attempt to show, as he might otherwise have done, that Cicero did not mean to deny the existence of a living, personal God, who governs the world.

"But," said he, "does not the book of Nature—your Bible, as you call it—have something to say of God? Does it not speak of an infinitely wise and good Creator and Governor? Do not the works of Nature tell of the same God whose being and character were preached to us yesterday from the Holy Scriptures?"

"Nature has never spoken to me of any God except herself. What need is there of a creator? Who can prove that the universe did not exist from eternity? Nature has her laws of development, and under those laws all the operations of nature go on. You had better read Darwin. If one must find the character of God in nature, he may as well picture an evil creator and governor as one that is good and righteous. Does Nature punish those whom you call the wicked? Does Nature reward the righteous? Do not the laws of Nature bring suffering to the good and the bad alike, and happiness also to all classes of men? Would you, if you had power, create a world like this—a world in which danger, pain, and death, in every shape, lie in ambush against its inhabitants every hour of their poor existence? But I must go." Pausing a moment, however, as if reluctant to go, with a voice sad and almost tremulous, which revealed a great deal more of his heart than he designed to express, he added: "God knows, deacon, if there be a God, how I wish I knew the truth about these matters. The world and myself are to me great and dreadful mysteries."

"'He that will do his will shall know of the

doctrine,'" answered Deacon Gregory; and inviting him to come to church again, they separated.

This conversation with the pious deacon, though he had himself done most of the talking and had his say almost unopposed, did not tend at all to bring rest to Mr. Hume's conscience. He saw that the deacon's faith in God did for him more than belief in Nature and worship at the altar of Science could do for unbelievers. He felt also that he had spoken a little too freely, especially in revealing, at the last, his unrest of spirit from the want of fixed convictions in regard to religious truth. Deacon Gregory, by the sincerity and manliness of his address, was accustomed to draw out the hearts of men so that they expressed them more freely than they designed.

Upon a bench in a shaded corner of the store sat a lad of sixteen or seventeen years, unnoticed for the time being by either Mr. Hume or Deacon Gregory. His name was Ansel, and he was the son of the senior deacon of the church. He was in the village academy, and had there been nearly fitted for college. He stood at the head of his class, and, with his sharp intelligence, his impetuous energy, and high ambition, every one was predicting for him a distinguished life. He had grown up thus far in the bosom of a family where piety was no pretence. Earnest prayer had gone up for him by day and by night. He had been well trained in the Sunday-school, and for a year had been a member of the small class of young men taught by Mr. Wilton. He had always shown a ready interest in all Bible studies and a quick understanding of Scripture doctrine, so that some thought him not far from the kingdom of God. But Deacon Arnold little thought what was in the heart of his son. He might have known, for to read his son's heart he had only to recall his own early manhood. For years he had hung trembling upon the brink of ruin, swept, at times, by his self-will and turbulent youthful passions, to the very verge of the precipice, and had been preserved only by singular grace from falling over. Now Ansel was following in his father's early footsteps— self-willed, and stubborn against the Spirit of God, and, at times, almost persuaded to cast off all religious restraint, that he might carve out his worldly fortunes untrammeled by religious or conscientious scruples. He had rarely heard infidel sentiments expressed, but the little that he had heard had attracted him, and had encouraged him to give loose reins to his own unbelieving disposition. It had not escaped his notice that the two or three men whom he had heard spoken of as infidels were among the most respectable and shrewdest business-men in the village. The idea, moreover, of rejecting all authoritative doctrine, and believing whatever should please him, carried with it so free and independent an air, and harmonized so well with his natural disposition, that he easily drifted in the direction of unbelief.

Sitting this evening unobserved, he drank in every word which Mr. Hume uttered. Some of the notions thrown out were quite new to him. "The book of Nature my Bible"—"Nature reveals no God but her own laws"—"No proof that the matter of the universe has not existed from eternity uncreated"—"Nature has her laws of development"—"No need of a God to govern the world,"—these were seed-thoughts in Ansel's mind. He had before thought of the only alternative to be set over against belief in the sacred Scriptures as simply unbelief—bare, blank denial of their truth. He had not dreamed of building up a set of proud, rationalistic notions, and denying the truths of religion in the character of a young philosopher. He kept his thoughts to himself, and turned them over and over in his mind during the week, and when again he met his pastor in the Bible class his head was full of his new notions. The lesson went on, however, and closed as usual. It so happened that this was the last in a series of lessons upon the Gospel of John. It was necessary, therefore, that another course of lessons should be decided upon.

Mr. Wilton proposed the question to the class: "What shall be our next course of lessons? Would you like to study one of the Epistles—the Epistle to the Romans or that to the Hebrews?" And he briefly stated the subject discussed in these Epistles of Paul. "Perhaps," he continued, "you would prefer to study one of the historic books of the Old Testament?" The class had no opinion. They wavered between an Epistle and a historic book and topical lessons which should confine them to no one book of the Bible. Then Ansel spoke up:

"Mr. Wilton, why can we not study something which we know to be true?"

Ansel meant to be very cautious as well as

very respectful, and did not design to commit himself by suggesting his own thoughts. He was respectful, but in the confusion of the moment he had brought out the very thoughts which he meant to conceal.

Mr. Wilton was startled, though he did not fully understand the drift of Ansel's question.

"What do you mean, Ansel?" he asked; "do you think Genesis less trustworthy than the Epistle of Paul?"

Ansel saw that he had committed himself and must now make the best of his situation. He therefore answered cautiously:

"Some persons, I have heard said, do not believe the Bible to be inspired, and they say that we have no evidence that it is true."

"What have you been reading, Ansel, that has put such thoughts into your mind?"

"I have never read a book that said anything against the Bible."

"But what did you mean? Do you wish to study the evidences of the truth and inspiration of the Holy Scriptures?"

"I should indeed like a course of lessons upon that subject, but that was not quite what I was thinking of."

"What book can you find which is true if the Bible is not true?"

"I do not know, sir, but I heard Mr. Hume say that the book of Nature is his Bible, and that we do not need any other, and that, whether the Bible be true or not, we know that the teachings of Nature must be true."

"But we should find," said Mr. Wilton, "that the teachings of Nature and the Bible would perfectly agree. Did Mr. Hume say that what he calls 'The book of Nature' contradicts the sacred Scriptures?"

Now that Ansel could give the thoughts which filled his mind, not as his own, but as Mr. Hume's, he showed no farther hesitation in speaking.

"Yes, sir," he answered; "he said that Nature teaches us that there is no God, because there is no need of any. He said that we cannot prove that God created the universe, but that matter has existed from eternity uncreated, and that all the changes in nature go on by certain laws of development, and that a certain Mr. Darwin had written a book and proved this."

The reader will notice that in the report of Mr. Hume's language the scholar went somewhat farther than his master had done. Mr. Wilton was well acquainted with the present shape of scientific infidelity, and saw that Ansel's statements were somewhat exaggerated, but he understood in a moment the drift of Ansel's thoughts, though he could not tell as yet how deep and fixed an impression had been made upon his mind. But he did not care to probe Ansel's conscience just then and there, in order to learn the exact state of the case.

"If I understand you, then," he said, "you would like a course of lessons in the teachings of Nature?"

"Of course, I did not suppose that you would allow us to have a course of lessons in the works of Nature instead of the Bible."

"But if I were willing to give you a course of lessons showing the footprints of the Creator, so to speak, in the physical world, how would it please you?"

"I should like it very much."

"How would such a plan please the other members of the class?"

The idea was entirely new; no one of them had ever dreamed of studying in a Bible class anything except the Bible; but young people are not averse to novelties, and they readily gave their assent. Yet I should do the class injustice by leaving the impression that they were influenced simply by the love of something new. They were of just that age when one hardly knows whether to call them lads or young men; they had been well instructed, and were just beginning to think independently. They were rapidly becoming conscious of their own mental power, and were eager to try their strength upon every line of thought. Their own weakness they had hardly begun to learn. Perhaps they were all the more ready to undertake such a course of study because they knew nothing of the difficulties attending it.

The tinkling of the superintendent's bell warned them to close their conversation.

"We have not time to-day," said Mr. Wilton, "to fix on the particular line of study which we shall follow. Of course we cannot examine all the works of Nature, and study every science, and trace the footprints of the Creator in every place where he has walked; we must fix on some small part of the works of God, and direct our attention closely to that. We shall find this course more profitable than roaming carelessly over a much larger space. Our next lesson will have to be a general one—a kind of preface to what shall come after. In the mean while, you can be collecting your thoughts upon the subject, and calling to mind anything that you have read bearing upon the handiwork of God manifest in Nature."

The school closed, and as the scholars pass out let me introduce to you the members of the pastor's class. This class was small for several reasons. The church to which Mr. Wilton preached was not the popular church. The fashionable people and all who loved popularity and drifted with the tide went to another church. Careless, thoughtless young people naturally went with the crowd, and of those who attended his church some did not care to join his class. He was too much in earnest to please them. He made religion a reality, and his instruction compelled them to think, and of course those who did not like to think were not well pleased with him. But there were a few of the young men who were greatly interested in his instructions. They were earnest readers of instructive books; they liked conversation which called out thought; they were most of all pleased with questions and themes which gave them new ideas. Indeed, in the community, there were two classes of persons who held Mr. Wilton in the highest esteem and regard: one of these was composed of men and women of earnest, intelligent piety, experienced Christians; the other, of those who were not Christians, but who respected sincerity and disinterested godliness, and liked sermons filled with meat and marrow.

Thus, at the present time, we find his class composed of but three young men. With Ansel you are already acquainted. The second is Peter Thornton, the son of a master-carpenter. He was frank, outspoken, quick in the acquisition of almost every kind of knowledge, but very little given to silent reflection. He listened to his pastor's instruction as he would go to a well-filled library, to draw out its stores of information. Morals and moralizing he did not like. He was not pious, and gave no indication of serious impressions. The third was Samuel Ledyard, the son of a poor widow. By painful industry and economy his pious mother was giving him the best advantages for education which the village afforded, praying the Lord to give him a part in the blessed work of preaching the gospel and winning sinners to Christ and salvation. When but twelve years of age he gave himself to Christ, and had been trying faithfully to follow his Lord. The long winter evenings were spent in reading books of history and science—books fitted to furnish and strengthen his mind—and long ere the light dimmed the morning star he was poring over his Bible, alternately reading the word and praying that his mind might be opened to understand the truth in its beauty and greatness, and that the truth might be wrought in him by vital experiences.

With such habits it was no wonder that he grew in grace—it was no wonder that he grew in all manly qualities. He was silent, meditative, and retiring, as gentle in his ways as a quiet girl, yet all who knew him recognized in him a singular weight and worth of character. Those to whom the Lord revealed his secrets began to say that Samuel was appointed of God to preach the gospel, and his mother felt the assurance growing strong in her heart that her prayer was granted, and that the Lord was preparing her only son and only child for a place in the gospel ministry. If only she might train up a son to such a work, and when she should go to her rest leave in her place a man working for Christ in his harvest-field, gathering sheaves unto everlasting life, she felt that her cup would be full. She was ready to say with Simeon: "Now lettest thou thy servant depart in peace, for mine eyes have seen thy salvation." How unlike she was to those mothers who lay all hindrances in the way of their sons entering the work of the Christian ministry, willing that they should do anything but this! and how different from those who declare that their daughters shall never wed ministers of the gospel, teaching them to despise the service of a pastor's wife! How often God gives over such sons and daughters—children consecrated from their birth to worldliness—to be entangled and lost in worldly snares! As such mothers sow, thus also do they reap.

These were the three lads, just growing into young manhood, at this time under the instruction of Mr. Wilton. He was not ashamed of his class, though it was small. As he saw them expanding in thought and taking shape under his hand, he felt that in them he was perpetuating his influence in coming generations. He believed that in one or more of them he should preach the gospel after his body was sleeping in the earth awaiting the resurrection.

I trust the kind reader will be interested in following the course of study through which their pastor shall lead them.

CHAPTER II.

NEW THOUGHTS FOR THE SCHOLARS.

THE little class which has been introduced to the reader came together the next Lord's Day interested and expectant, yet not knowing what to expect. They had chosen a course of study, yet they could not tell what that course was to be. They had tried to think of something definite about it, but could fix their minds upon nothing. In fact, the whole subject was new, and they could not decide where or how to take hold of it. They came together, therefore, with no more knowledge of the subject than when they separated.

Mr. Wilton himself came before his class in a state of doubt. He had given the subject many hours of thought, and had carried it to his closet and besought the guidance of the Holy Spirit, for he believed the divine Spirit to be the best guide in understanding the works as well as the word of God. He felt that his prayer had been heard and answered. He was prepared, therefore, to speak with the force of clear understanding and positive convictions. But the precise line of study he had left to be determined by circumstances, perhaps by the previous studies of his class in their academic course. This was to be decided by further consultation.

"Since no lesson was assigned upon which you could prepare yourselves," Mr. Wilton said, after the opening exercises of the school were finished, "I shall spend the half hour to-day in a kind of conversational lecture. You may call this the preface or introduction to the lessons which will follow. I shall try to make plain some general principles which we must keep in mind, whatever department of God's works we shall attempt to examine. I wish you to feel entirely free to interrupt me at any time, and ask any question or present any objection which may strike your minds. We must, if possible, have no prowling bands of enemies in the rear. I wish to make everything as plain as the case will admit.

"One thing let me remind you of in the beginning: I shall not try to prove to you that there is a God. I shall not try to prove that the world had a creator. There are some things which men do not believe merely on account of good evidence, nor disbelieve for want of proof. Men believe in their own existence, but not from a course of argument. Most men believe in the real existence of the outward world—the earth, the hills, the rivers, the trees, everything which we see and hear and feel—but not on account of proof. Here and there a strange man is found who professes to disbelieve the real existence of all material things, but he disbelieves not for want of proof. Men believe that their sight and hearing and touch do not deceive them, but their confidence in them is not the result of a course of reasoning. To believe in our own existence, and in the existence of the world outside of us, and in the truthfulness of our senses, is natural; to disbelieve these things is unnatural: it shows a state of disordered mental action. When such disbelief is not practically corrected by a man's understanding he is counted insane, and is treated accordingly.

"Belief in the existence of God is also a natural belief. A denial of God's existence shows, not disordered mental action, but a disordered moral and spiritual state. It shows the absence of that spiritual faculty by which we receive spiritual impressions, and are brought into contact with the spiritual world, and hold intercourse with God and Christ and the Holy Spirit. Men must be convinced of the existence of God through their conscience, their moral and spiritual nature. Do not misunderstand me. I do not say that good evidence cannot be brought to prove to one's reason the existence of God, but God has not left his existence to be *proved:* he has *revealed* himself to men's consciences and to their faith; and those in whom conscience and faith are well developed, sound, and right do not need an elaborate argument to prove the divine existence. I shall simply try to show that the works of creation exhibit the wisdom and goodness of God. If any man, looking at such indications of wisdom and kindness, can believe that it all comes by chance or is the work of some evil agency, and that no Being of boundless intelligence, wisdom, power, and goodness has anything to do with the making and governing the world, he certainly shows great prejudice: he does not want to recognize God's existence. He must be one of those spoken of by the Psalmist who say, 'no God.'

"During my recent visit to Greenville I visited a mill, the largest of its kind in the country. In one room was a machine, something like a huge straw-cutter, working with great power. In another room was a large steam boiler hung upon a shaft and made slowly to revolve while filled with steam. In a third room were large oval tanks, or cisterns, which might be filled with water. Across each tank was a heavy shaft carrying a drum set with steel blades, and as the drum revolved these blades passed other blades in the bottom of the tank, cutting whatever came between like scissors. In a fourth room were certain long and complicated machines. Each machine was composed mostly of rollers. There were large rollers and small rollers, solid rollers of enormous weight, and hollow rollers to be heated by steam within. Over and around a portion of these rollers passed a broad wirecloth belt. Over others passed a like belt of felted cloth. With these machines before you, could you tell me whether the inventor were a wise and skillful machinist?"

"How could we tell," asked Peter, "without knowing what kind of work the machine was designed to do?"

"You could not tell," answered Mr. Wilton; "you would need to know both what the machine was designed to do and all the processes by which the work was to be carried on. This brings out the first point which I wish you to fix in mind. It is this: To judge of the wisdom of any contrivance, we must understand the purpose, or object, which the inventor had in view; we must understand the work to be accomplished, and also the difficulties to be overcome. An ordinary locomotive steam-engine is admirably fitted to run on iron rails, but he would be a foolish man who should purchase such an engine to draw a train of loaded wagons over a common road of earth. On such a road it could not even move itself. It is good for that for which it was made, and for nothing else. How would you apply this principle to the subject we are now considering? You may answer, Samuel."

"I think you mean," said Samuel, "that, in order to judge of the wisdom and goodness of God in creating and governing this world, we must know the object he had in view in making such a world."

"That is my meaning, and I am glad that you understand me so perfectly. If this world were created with no other object than to be the grazing-field for herds of cattle, which see no difference between the beauty of the violet and the dull shapelessness of the cold earth upon which it grows, and never lift their eyes above the horizon, then all the beauty of earth and sky would be useless; there would be no wisdom or goodness in the creation of this beauty. There would be no wisdom or goodness in laying up in store beds of coal, buried deep beneath the surface of the earth, if God designed the world to be inhabited only by savages too rude and ignorant ever to mine it, and turn it to some practical use.

"But let me give you another illustration, which can better be applied to the condition of things in this world. Just in the outskirts of one of our inland cities I once saw a large and elegant building, whether a private dwelling or a public institution I could not at first tell. It stood high and airy, commanding the most pleasing prospect that all the region presented.

We will follow a visitor as he goes to examine that noble establishment.

"As he comes nearer, he sees that the edifice is simple and classic in its style and chaste in its architectural adornment. It is a pleasure for the eye to rest upon its graceful symmetry. But in place of the light and graceful fence which he expects to find enclosing its grounds, he sees a stockade strong and high. The janitor turns the heavy key, the rusty bolt flies back, and the visitor enters the enclosure. Within the stockade he finds a portion of the ground laid out with taste and cultivated with choice and beautiful flowers; another part is devoted to the culture of garden vegetables. He finds workshops also for the manufacture of pails and tubs, brooms and mattresses. The visitor is ushered into the mansion itself. He finds everything more than comfortable; the rooms are heated from furnaces below; every part is perfectly ventilated; the windows command a view of the country around which must please the most cultivated eye; a school-room is provided with all needed apparatus for the most thorough instruction. 'Surely,' says the visitor, 'the founder of this institution must have been both wise and good.

He must have loved the young in order to study and supply all their needs so completely.' But some things strike the visitor painfully. The windows are grated with iron, and some of the rooms are almost like prison cells. 'Can it be possible,' he thinks within himself, 'that the young need to be confined by a stockade in so pleasant a place and shut in by grates of iron for the enjoyment of such advantages?' The master as he teaches his pupils seems as kind and gentle as a mother, yet there is a firmness and authority in his tones and a rigidity in his training, as if his government were kept braced against a mutinous spirit. The means of punishment also are provided, and, when occasion requires, stern chastisement is employed. All this seems to the visitor like an enigma. The institution appears to him like a bundle of contradictions. A father could not have provided a pleasanter home or larger advantages for his children, but fathers do not commonly surround their homes with stockades, and cover their windows with bars of iron, and train their obedient children with a hand of such firm, unyielding force. 'Pray, sir,' he says to the master, 'what is this strange contradictory institution?' 'It is the State Reform School,' the master answers. 'And who are these lads and young men for whom all this work and wisdom is expended?' 'They are those who have taken the first steps in crime, but have not as yet become hardened and fixed in wickedness, and are sent here with the hope of overcoming their vicious propensities and training them to virtue and an honorable manhood.'

"Everything is now made plain. The need of the stockade, and the grated windows, and the rigid government, as well as of the pure air, the garniture of beauty, and the kind loving care, is manifest. It is a place unsuited to a family of obedient children, and equally unsuitable as a place of confinement for confirmed criminals, shut up, not for reform, but for punishment. It is wisely adapted to the work designed to be accomplished, and to no other.

"In like manner, if we would judge of the wisdom and goodness of God in the creation and government of this world, we must understand the use for which the world was designed. Is this plain to you, Ansel, and does it seem reasonable?"

"Yes, sir; I think I understand it, and I can see no objection to the principle. I think even Mr. Hume could find no fault with that. But how shall we know the object for which God made and governs the world?"

"That is the next point to be considered. Perhaps you will tell us what seems to you to be that object? Young people sometimes have thoughts and opinions upon the greatest questions."

"I have never formed an opinion of my own," Ansel replied, "but I have always heard it said that God designed to show how perfect and good and beautiful a world he could make. But many things in the world seem to me neither perfect, nor good, nor beautiful."

"Why, Ansel!" exclaimed Samuel; "the Bible says that 'God saw everything that he had made, and behold it was very good.'"

"And, Mr. Wilton," asked Peter, "does not the Bible say that 'God created all things for his own glory'?"

"Before answering any of these questions, let me ask Samuel a question. What do you understand to be the meaning of the words you quoted from the last verse of the first chapter of

Genesis?—'God saw everything that he had made, and behold it was very good.'"

"I suppose it means," answered Samuel, "that God made everything just as good and beautiful as it can be, so that any change must be a change for the worse. The lecturer last winter said that if men could entirely destroy any one of the most troublesome species of insects, their destruction would be a great loss to the world, and that if a single atom of matter belonging to the earth were annihilated, it might throw the solar system out of balance, so that it would finally be destroyed."

"I remember," said Mr. Wilton, "that some lecturer last winter made statements of that kind, and I have heard other people declare that the least possible change in the world would be injurious, if not destructive, to the interests of man, and that the most troublesome beasts and insects and the most loathsome reptiles are necessary to human happiness. Does that seem to you to be true, Samuel?"

"I have always tried to believe it, because I thought I ought to believe it. It has seemed to me to be dishonoring God to believe that he did not make the best possible world."

"You are right in trying to believe what seems to be right and true, even though difficulties do lie in the way. Difficulties do not by any means show that an opinion is false. We must certainly believe that God made this world perfect for the object which he had in view in making it. But not a few skeptics deny the existence of a good, wise, righteous Creator and Governor, because they have a wrong idea of the end for which the world was created, and, consequently, a wrong idea of that in which its perfection must consist. Let me ask you a few questions which will lead your minds in the right direction. Do not men produce by cultivation better fruits and vegetables than Nature ever grows when left to herself?"

"Yes, sir," said Ansel; "the peach and apple and potato have been brought up to their present state of excellence by great care and exertion. Originally, they were almost worthless."

"And not only that," said Mr. Wilton, "but when once that careful culture is relaxed they begin to return to their former badness. Again, do we not improve upon Nature by drainage and improve upon the climate by irrigation?—in fact, do not men by drainage and irrigation and all manner of culture greatly improve the natural climate of a country?"

"I think that is true," said Ansel.

"I never thought of that before," said Peter.

"Moreover, do you not suppose that heaven will be more beautiful than the earth, and that a thousand troublesome things besides sin—loathsome sights, discordant and jarring noises, disgusting and nauseous odors—will be absent from that 'better land'?"

"And *I* never thought of that before," said Samuel. "I am sure that many unpleasant things besides those which sin has brought into the world will not be found in heaven. I see that this world might be changed and not be made worse for holy beings to live in."

"The world is very good," said Mr. Wilton, "for the purpose for which it was created, but we need not look upon it as designed for a specimen of the most beautiful, pleasant, and desirable world which the Creator could produce."

"But you have not told us," said Peter, "what the Bible means when it says that God created all things for his own glory. Does it not mean that he made the world so good and perfect that all creatures ought to praise him on account of it?"

"We ought," said Mr. Wilton, "to praise God for the wisdom and goodness displayed in the works of creation. That is the teaching of the Bible in many places; it is also the sentiment of the Bible that God created the world and carries on all things for his own glory, but it nowhere uses the exact language which you have employed. In Isa. xliii. 7, speaking of 'every one that is called by my name,' the Lord says, 'I have created him for my glory.' In Prov. xvi. 4 it is written, 'The Lord hath made all things for himself; yea, even the wicked for the day of evil;' and the four and twenty elders fell before the throne of God saying: 'Thou art worthy, O Lord, to receive glory and honor and power; for thou hast created all things, and for thy pleasure they are'—that is, exist—'and were created.' I might quote other texts of similar meaning. We are taught also that our first and supreme aim in all our conduct should be the glory of God. 'Whatever ye do, do it all to the glory of God.' But here two questions arise: What is the glory of God? and, What is it for God to glorify himself by his works of creation and government? Who will tell us?"

All were silent, and Mr. Wilton went on speaking: "The word glory means, first and literally, a halo of light. The glory of God is the radiance, or halo, so to speak, of his infinite attributes and holy character. God glorifies himself when he reveals himself, and makes known his character, and causes the uncreated splendor of his attributes to break forth, so that his creatures recognize them and adore him. This, you see, is very different from the idea of glory among ambitious men. God glorified himself in the creation of the physical world, because from that creation his wisdom, power, and goodness are manifest. He glorified himself in the creation of angels and men, because they were created in the image of God and are finite pictures, so to speak, of the infinite Creator —a revelation of his spiritual being and personality. He glorifies himself in his government of the world, because his administration of affairs exhibits his justice, mercy, and holiness. This is what we mean by the glory of God and his working all things for his own glory. This is somewhat difficult for persons of your age, so we will leave it and return to the exact subject of discussion. Admitting that God created the world and governs it for his own glory—that is, to reveal himself—for what specific purpose did he design this earth?"

"I don't know," said Peter, "that we understand what you mean by 'specific purpose.'"

"Very well, then," said Mr. Wilton; "I will suggest the answer. Does the world seem as if fitted up to be the dwelling-place of holy beings?"

"I have never thought of the question before," said Ansel; "but it seems to me that many things in this world would give pain even to angels if they lived here with bodies like ours."

"I agree with you, Ansel. If men were sinless and holy as the angels of heaven, many things in this world would bring them distress. But does it seem reasonable that the world was designed merely as a place of punishment for men by reason of their wickedness?"

"Some men are not wicked," replied Samuel. "There have always been men willing to die rather than disobey God. Surely, God does not punish such men. And many beautiful and pleasant things are found in the world—arrangements plainly designed for the welfare and happiness of men."

"I think you are right, Samuel. But, without asking further questions, I will give you the conclusions to which my study upon this subject has brought me, and some of the reasons for those conclusions.

"This world was made chiefly as the dwelling-place of man. The world was not planned merely as the abode of brute animals. Men are nobler than the brutes. Men have permanent interests and advantages. Aside from the glory of God, men are an end unto themselves. To become and be *men* is the noblest object of human life, but the animal tribes exist for the use and benefit of others. To be an end to itself, a creature must be immortal; but the brutes exist for the use and advantage of man, live out their transient life, and exist no more. This is the view presented in the sacred Scriptures. God gave to man lordship over the earth—not only over the soil to subdue it, and over the great forces of Nature to bring them into subjection for human advantage, but also over the brute creation, 'over the fish of the sea, the fowls of the air, and every living thing that moveth upon the earth.' I conclude also that God did not prepare this world as a prison-house and place of punishment for rebels against his government. Too many pleasant things abound for me to believe that. The pleasant breezy air, the glorious sunlight, the refreshing showers, the treasures of mineral wealth stored up in the earth, the fertile land and golden wheat, the beauty spread over all nature, the sweet consciousness of existence, so that just to live and act is joy, and the comfort and hope of immortal pleasure enjoyed by truly Christian men,—all these things, and many more, assure me that not the subtle shrewdness of a tormentor nor the unmingled justice of an inexorable judge, but the heart of a kind and loving Father, planned our earthly dwelling-place. You said, Samuel, with truth, that there are many pious men in the world who are dear to God, and Paul says, 'We know that all things work together for good to them that love God.' For those dear ones Christ has such love that he counts everything—whether good or bad—that is done to them as if done to himself. 'Inasmuch,' he says, 'as ye have done it unto one of the least of these my brethren, ye have done it unto me.' Moreover, Jesus said: 'For God so loved the world, that he sent his only begotten Son, that whosoever believeth in him should not perish, but have everlasting life.' From these words of Jesus we see that there is love manifested in the dealings of God with the inhabitants of our world. Were it not so, there would nothing remain but a 'fearful looking-for of judgment and fiery indignation, which shall devour the adversaries.'

"On the other hand, I conclude that God made the world as the dwelling-place, not of obedient, holy children, but of those who are disobedient, fallen, and alienated. These disobedient and alienated ones he holds under discipline and chastisement, in order to keep their wickedness in check, to recover them from their sins, and train them up in virtue and holiness, or to remove from the obstinate and incorrigible all excuse for their sins and all plea against their final condemnation. In doing this he glorifies himself by manifesting his wisdom, goodness, mercy, and holiness.

"This opinion seems probable from the fact that this is the purpose for which God has actually used and is now using the world. Here he keeps and governs the human race. This race is made up neither of holy beings nor of hopeless reprobates. They are the creatures of God; fallen indeed, yet loved; sinful, but objects of divine compassion; deserving of righteous wrath, but the recipients of the offers of salvation through Christ. Even penitent believers in Christ and devoted servants of God are not free from evil propensities, but need to be kept under constant training and discipline. This is the use to which the Creator has actually put the world. Is it not reasonable to believe that he designed it for their use? Ought we to believe that God planned the world for an object for which it never has been and never will be employed?

"If sin were removed from the world, the chief part of human suffering would be removed. This no man can deny. Wars would cease; the want, disease, and woe resulting from selfishness, idleness, and vice would disappear, and nothing would stand between man and his Maker. What new life and joy would fill the world if free communication were restored between man and God, and the divine smile were again to enlighten the world! It would seem that heaven had enlarged her borders to embrace this earthly ball. But the fact would still remain that this physical world is unfitted to be the dwelling-place of sinless beings. The constitution of the world would bring upon them pains and evils which would seem a most unworthy heritage for loving and obedient children of our heavenly Father. Let sin be taken away, and wearisome toil in subduing the earth would remain. The soil of the earth is hard and clogged with stones, and clammy with stagnant waters, and sown well with the seeds of noxious weeds, and overgrown with thorns and thistles. Endless watchfulness and toil is the price of a livelihood. With the sweat of his face man must eat his bread. An army of enemies have pre-empted the soil which man must till. This state of things the word of God refers to sin: 'Cursed is the ground for thy sake; in sorrow shalt thou eat of it all the days of thy life. Thorns also and thistles shall it bring forth to thee.' The necessity of toiling as we do now for our daily bread, God denounced upon man as a curse on account of sin. We cannot, therefore, regard this as a suitable condition for sinless beings.

"This burden of toil is lightened by the progress of modern sciences and inventions much less than some men think. Every step of progress has been made by the sacrifice of hecatombs of human lives. From our laboratories and workshops products of human skill, rich and rare, are sent forth; but what are they but smelted and hammered and graven and woven human bones and sinews, the health and life of men? No means have been discovered by which the most necessary processes of the arts can be made otherwise than dangerous to health. Only when thousands of miserable workmen had perished was Sir Humphrey Davy's safety-lamp invented; and now the danger, to say nothing of the hard toil, of the collier's life is only lessened, but not removed. Still, our furnaces roar and the whole tide of civilization goes on by the health-destroying servitude of men, buried alive as it were in the dark bosom of the earth. Would that seem to be a fitting employment for the sinless children of the all-loving Father? Employés in many kinds of manufacture slowly sink under the accumulated evils of daily toil, and no means of making their employments healthful have been discovered.

The friction-match, which has become so nearly a necessity, is made by a process so destructive to health that only a certain class of laborers can be prevailed upon to do the work. I might go on to speak of other painful circumstances in which men find themselves by the almost antagonistic attitude of Nature. But if we reject these dangerous processes of manufacture and art, we go back at once to the wooden plough, the distaff and tinder-box of primitive times, and also to primitive poverty and primitive toil, and, I may also add, to primitive exposure to the hostile and pitiless forces and inclemencies of Nature. Purge the earth of sin, and wearisome toil would still remain. Nature must be nursed and cultivated or she yields no bread. Her hostile attitude must be overcome; the thorns and thistles must be rooted out; and every step of progress, won by suffering, must be held by painful work and watchfulness; otherwise Nature returns to the wild and savage state. Relax the culture of the choicest fruit, and it begins to deteriorate; leave the best-blooded breed of cattle to itself, and it returns again to the level of native, uncultured stock.

"The inhabitants of this world are also liable

at all times to diseases and destructive accidents. This condition of things could not be changed without changing the entire structure and plan of the world. Is that a fit dwelling-place for a sinless being where chilling winds one day shrivel his skin and fill his bones with rheumatic pains, and the next, sweltering heats pervade all his system with languid lassitude—where miasma lies in wait unseen to poison his blood, kindle the malignant fever, and bring him to the shades of death, and every form of accident crouches in ambush, ready to spring upon his victim unawares and tear him limb from limb? We cannot see that the absence of sin would dissipate this liability to disease and the danger of accidents. Nay, this liability and danger are written upon the very constitution of the human body. The finger of God has engraved it upon every muscle and bone and life-cell. The Creator gave the body that wonderful power called the *vis medicatrix*—the power of recovering from injuries and repairing damage done to itself. Pull a leg from a grasshopper and another grows in its place. By this we know that the Creator understood the liability of this little insect to lose a limb, and prepared him for it. In like manner the power in man's body to heal a wound or join a broken bone gives us to understand that the Creator expected man to live in the midst of danger. The precaution proves the risk.

"These accidents are such as no possible carefulness could guard against. To say nothing of the fact that all our knowledge of these perils comes from a painful experience of danger and death, what care, even after ages of sad experience, could ward off the thunderbolt? What carefulness could guard against the tornado on the land, or the hurricane and the cyclone upon the sea? Who should stand sentinel against the unseen poison borne upon the wings of the wind? What power should save him from the bursting of the volcano and the jaws of the earthquake? What care could give him knowledge of the qualities of all natural substances, that he might avoid their dangerous properties? We can suppose a divine care over man that should do all this and save men from harm, but it would be a providence superseding all human knowledge and exertion—it must be a providence to which the human race is now a stranger; miracles would then be the rule, and the undisturbed course of Nature the exception.

"If, however, we suppose that God designed the world as a training-school, so to speak, of fallen beings, such as the word of God declares the human race to be, all is plain, everything is suitable and harmonious. We can see the fitness of at least the chief outlines of man's earthly condition, and can perceive God's wisdom and goodness in the constitution of the world.

"The pain and woe-producing agencies of Nature are seen to be not at all contradictory to goodness, but on the other hand eminently wise and righteous. The whole sum of human misery expresses God's displeasure at sin. By their sufferings men learn how abhorrent is sin in God's sight. By the consequences of evil-doing they learn not to transgress. As none are free from the taint of depravity, none are free from pains. The necessity of labor—one of the elements of the primal curse—is a check to sin on the part of the vicious, and a discipline and trial to virtue on the part of the penitent. The multiform trials of life—which can indeed be borne well only by the grace of God—while they teach the evil of sin and keep the heart chastened and subdued, nourish heroic and dauntless virtue in the faithful. 'Daily cares' become 'a heavenly discipline.' Dangers and calamities startle the stupid conscience, and keep alive the sense of responsibility to God on the part of the wicked; they quicken the sense of weakness and dependence in the believing and educate their faith in God. The more sudden and overwhelming these evils, and the more these dangers are placed beyond the possibility of being warded off by human care, the more do they awaken in men a sense of the divine presence and of responsibility to God.

"But would not all these natural agencies subserve essentially the same ends in the discipline of unfallen and sinless beings? By no means. If sufferings came upon a sinless being, he could not feel that they came as chastisements; he could not feel them to be deserved. They would be to him a 'curse causeless,' and hence would bring no advantage. He could only cry out in astonishment, 'Father, why am I, thine obedient son, thus smitten?' Calamity falling upon the innocent would be an anomaly in the universe. But now the sufferer, pierced through and through with a sense of ill desert, meekly bows his head, murmuring, 'Father, all thy judgments are just and right.'

"One very important feature of the world we live in is its moral symbolism. The world is full of most suggestive symbols and emblems of moral good and evil. There are all beautiful and glorious things, to stand as types of goodness, truth, and righteousness; there are all loathsome, malignant, and hideous things, to serve as the types of folly and wickedness. Was it merely an accident that the dove was fitted to become the emblem of purity and of the Holy Spirit? the lamb, to be the emblem of gentleness, of Christ the gentle Sufferer, and of his suffering people? the ant, to be the type of prudent industry? the horse, of spirit and daring? and the lion, of strength and regal state? Was it only an accident that prepared cruel beasts and disgusting, poisonous reptiles as the types of evil passions and sins—that made the venom of the viper, the cunning of the fox, the blood-thirstiness of the wolf, the folly of the ape, and the filth of the swine, symbols of foul, subtle, malignant sin and folly? Nature is full of these emblems. The palm tree with its crown of glory, the cedar of Lebanon, the fading flower and withering grass, the early dew and the morning mist, the thorn hidden among the leaves of the fragrant rose, poisons sweet to the taste, and medicines bitter as gall,—how all these natural things preach to men sermons concerning spiritual verities! There is no virtue or grace which is not commended to man by its image of beauty in the animal tribes; there is no vice against which men are not warned by its loathsome, disgusting form shadowed out in the instinctive baseness of irresponsible brutes.

"Thus we find earth, air, and sky to be full of silent voices proclaiming in the ears of man that which he most of all needs to remember. These types and symbols of virtue and vice are specially needed by fallen beings. They seem fitted for beings whose spiritual eyes are blinded and all their spiritual senses blunted—beings with whom there is no longer 'open vision' of spiritual realities. These pictures of evil are most impressive to men who see in them the reflection of their own base passions. How the fetid goat and the swine wallowing in the mire speak to the lecherous man and the drunkard! In a world of sinless beings these mimic vices would seem rather to mar God's handiwork.

"Set the human race, fallen as it is, in a

world where the patience of daily industrious toil would not be needed, and the race would rot with putrid, festering vice. Remove all danger, and men would forget and deny that the Creator holds them responsible. Let no evil consequences follow evil-doing, and men would cease to make a distinction between right and wrong. Take away death, and they would deny the existence of a spiritual world. But in this world God has hedged men around with checks and penalties and painful discipline, such as are of use only in dealing with sinners.

"I conclude, therefore, that God prepared this world as it now is as a place of discipline for a fallen race. This is the use to which he has devoted it in the past; and when there is no longer need of such a world for the discipline of men, we learn from the word of God that a 'new heaven and a new earth' shall be provided. This world is thus declared to be an unfit abode for the glorified saints. To judge, then, of the wisdom and goodness of God in the works of nature, we must keep in mind the object for which the Creator prepared the world. Ansel, tell us how this strikes you."

"I never thought of it in this way before," he

answered; "indeed I have thought very little of this subject, but—" Tinkle, tinkle went the bell upon the superintendent's desk. This was the second time the superintendent had struck his bell, but Mr. Wilton had been so intent upon his subject that he did not hear the first ringing.

The school was dismissed, but Mr. Wilton remained with his class to fix upon the particular department of nature which they would study. He found that all were studying natural philosophy, and had recently gone over the subject of heat. At his recommendation, therefore, they agreed to examine, as a specimen of God's works, his management of heat in the world. Mr. Wilton requested them to review the subject during the week, and be prepared to state and apply the general principles touching the nature, phenomena, and laws of heat which they had already learned. This work they will enter upon next Lord's Day.

CHAPTER III.

A DIFFICULT QUESTION.

DURING the week, Ansel, Peter, and Samuel were busy reviewing and fixing in memory what they had already learned of the nature and laws of heat. They were not only interested in the new line of study, and desirous of pleasing Mr. Wilton, but they also felt that their scholarship was to be tested, and each one was ambitious of standing equal to the best.

Ansel, of course, was busy and ambitious. The lesson was coming somewhat upon his own ground, and he felt in no wise unwilling to show how well he had mastered the subject. He entered upon it with feelings a little different, however, from his anticipations. The explanation which Mr. Wilton had given of the purpose of the Creator in making such a world seemed to him very reasonable. He could make no objection to it. But that explanation had taken away at one sweep a whole store of objections to God's goodness which he was waiting to bring out as soon as a good opportunity was presented. A world designed for the dwelling-place of sinners—sinners not already given over and doomed to final wrath, but to be recovered from sin and trained in virtue and holiness, or, if incorrigible, to be held in check and used as helps in the discipline of the righteous—he plainly saw must be as unlike a world fitted up for holy beings as a reform school is different from a home for kind and obedient children. Those arrangements which he had thought the most painful and objectionable might, after all, be the wisest and best. He did not see where to put in a reasonable objection to Mr. Wilton's unexpected argument, yet he did not feel quite satisfied to confess to himself that he was so soon and so easily defeated.

In this state of mind, on Saturday morning he met Mr. Hume upon the street.

"Good-morning, Ansel," said Mr. Hume.

"Good-morning," returned Ansel.

"I hear," said Mr. Hume, "that you have

given up studying the Bible in your Bible class, and have begun the study of natural philosophy. Is that so?"

"Not quite true, Mr. Hume. We are to examine some department of the works of Nature, and see what indications appear of the Creator's wisdom and goodness."

"That is a little different from the report which came to me. But what did you learn last Sunday?"

"Mr. Wilton told us that in order to judge of the wisdom and goodness of God in any of the affairs of this world we must consider the object for which that arrangement was designed. He said that if a man examine a cotton-gin, supposing it to be a threshing-machine, he would be likely to pronounce it a foolish and worthless contrivance; and that the fine edge of a razor would be worse than useless upon the cutter of a breaking-up plough. He told us that the earth was not prepared as the dwelling-place of sinless beings, but as a place of discipline for the fallen human race, and that we ought not to look upon it as the choicest specimen of workmanship which the Creator could construct."

"I have heard that Mr. Wilton believes something of that kind. Ansel, have you studied geology?"

"I have read a little upon that subject and have heard some lectures."

"Can you tell me, then, whether or not the natural laws which prevailed on the earth ages and ages ago, before the earth was fit for men to live upon it, are the same as those which have been in operation in these later ages, since men have inhabited it?"

"I suppose that the same laws have prevailed from the beginning of the geologic periods. I think that geology makes that very evident."

"If that were not so," said Mr. Hume, "the past history of the globe would be a riddle to us; it would be confusion worse confounded. In regard to those early ages we could not reason from cause to effect, for we should know nothing of the forces and principles then in existence. In geologic studies we judge the past from the present, and if that be not a trustworthy method of reasoning, all the conclusions of geologists are as worthless as dreams. Have you any reason to suppose, from what you have read on this subject, that a curse changed the character of the earth as a dwelling-place for man some six thousand years ago? Is it true, as Milton says, that then

> 'The sun
> Had *first* his precept so to move, so shine,
> As might affect the earth with cold and heat
> Scarce tolerable, and from the north call
> Decrepit winter—from the south to bring
> Solstitial summer's heat'?

Did the Creator then

> 'Bid his angels turn askance
> The poles of earth twice ten degrees and more
> From the sun's axle'?

Or was death then first introduced among the brute creation, as Milton fancies?—

> 'But Discord first,
> Daughter of sin, among the irrational
> Death introduced through fierce antipathy;
> Beast now with beast 'gan war, and fowl with fowl,
> And fish with fish; to graze the herb all leaving,
> Devoured each other.'"

"Animals must have died," said Ansel, "for their remains lie imbedded in rock which certainly existed before man lived on the earth."

"I wish you would ask Mr. Wilton one question for me."

"I am willing to ask him any proper question,

and I suppose you would not wish me to ask any other."

"I certainly would not. Will you ask him how it was possible for man not to sin and fall if God created the world for a sinful race myriads of ages before man was brought into existence? It would seem that if man had remained obedient he could not have lived pleasantly in a world prepared for sinners, and at the same time, by man's obedience, all the Creator's plans touching this world would have been dislocated and disappointed."

"I will ask him, sir," said Ansel, "at the first good opportunity."

This good opportunity occurred sooner than Ansel expected, for, before entering upon the proposed lesson the next Lord's Day, Mr. Wilton said to the class:

"I wish in these lessons to advance carefully and safely, and, as far as possible, have everything well understood. For that reason I invite you to speak freely of any difficulties or objections which may suggest themselves to your own minds or which you may hear presented by others. At the close of the last lesson the views which I had presented to you seemed very reasonable, but it is possible that, as you have thought upon the subject during the week, objections may have arisen in your minds. If so, I should be glad to hear them now."

"There are many things," said Peter, "of which I cannot see the use, even if we suppose that the earth was designed as the dwelling-place of sinners."

"It would be very surprising indeed if you could unravel all the mysteries of creation in a week's time. Wiser men than any of us have spent a lifetime in searching out the meaning of God's works, and died still in the dark upon many points. We need not expect to unravel and understand all the deep, complex, and delicately-interwoven contrivances in a world so vast and curious as this. The world is a great mystery—mysterious as a whole, and mysterious in all its parts—upon any supposition. But the explanation which I gave of its design furnishes a sufficient reason for the great outline of creation. This gives a reason for the pains and miseries which dog man at every step. This gives a reason for the earth's being left rugged and sluggish, bringing forth thorns and thistles, and requiring to be subdued by patient industry. It

CURIOSITIES OF HEAT.

shows a ground for the necessity of exhausting toil under a frowning sky and mid miasmatic airs—for the liability to diseases and accidents, and the hard necessity of death. These great elements of divine providence are not stripped of their halo of mystery, but with this explanation they are seen to form a harmonious whole for the accomplishment of a great and glorious purpose."

Mr. Wilton paused. Then Ansel said, "Mr. Hume wished me to ask you a question."

"Very well, I should be glad to hear it. I hope, indeed, that he sends his question from interest in the subject, and not with the design of perplexing us. I wish also that he were here to ask the question and hear the answer for himself. But what is the question?"

"He wished me to ask how it was possible for man not to sin and fall if God placed him in a world prepared for a race of sinners and unfitted for a sinless race. He said that in such a case, if man had remained obedient, the plans of God would have been disarranged."

"What answer did you try to give him, Ansel?"

"I did not try to make any explanation. It

seemed to me a very great objection. I did not see how such a course was consistent with God's righteousness."

"And you are not the first person who has objected to this as a great inconsistency. I am afraid the discussion will take more time than we ought to spare, but now that the question has been asked and the objection presented, I must take time to answer it, even if it consume the whole half hour.

"In considering this subject, as well as many others, we need to remember that the existence of difficulties is no objection to a principle or a fact. Difficulties wholly inexplicable by man attend facts and principles which must be true. A fact may be incomprehensible, though undeniable. The great Doctor Johnson said, 'There are insuperable objections against a plenum, and insuperable objections against a vacuum, yet one of these must be true.' What did he mean by that, Samuel?"

"He meant, I suppose, that we could not explain the possibility that any space should be wholly empty of matter, and could no more explain the possibility that any space should be filled with matter, but that all space must be filled, or else there must be empty space. Whether we can explain the possibility or not, one of them must be true."

"That is right. The same is true of many other facts besides a plenum and a vacuum. We cannot conceive of infinite space; we cannot conceive that space should not be infinite, but bounded. We cannot conceive of the creation of the world from nothing, and no more can we conceive of its eternal existence. The truth is that the mind of man cannot grasp such subjects so as to reason upon them correctly. No sooner do we attempt to reason about the infinite things of God than we run into absurdities and reach the most contradictory conclusions. And in this respect it makes no difference with what principle or proposition we start if it only contain some infinite element. Let me give you a simple illustration from geometry—an illustration which, very likely, is familiar to you: the larger a circle, the less is the curvature of the line which bounds it; that is, the more nearly does that line approach a straight line. An infinite circle must be bounded by a straight line, because with any degree of curvature the circle would be less than infinite. But a straight line cannot bound a circle. The attempt to reason about an infinite circle brings us at once to the most palpable absurdities and contradictions. Or take this illustration: the whole of a thing is greater than any of its parts. But divide a line of infinite length in the middle, and each part is infinite. We reach the conclusion either that the half is equal to the whole or that other wholly incomprehensible proposition, that one infinity is twice as great as another infinity. I have made these statements to show you that the existence of difficulties does not indicate, much less prove, that a fact is not real and true.

"Mr. Hume thinks the fact that the earth existed in its present condition before men sinned an insuperable objection to the view that this world was prepared as a place for the discipline of a fallen race. But let us look at the other side, and see if equal objections do not exist. The Creator foresaw the fall of man; is there no objection to the supposition that, knowing that man would sin, God made no provision for it? On the one supposition he foresees the evil and makes no provision; on the other, he foresees it and provides for the catastrophe. The former supposition certainly involves the greater difficulties.

"The objector may reply that the plan of God, by embracing the fall of man and including it as one of its essential elements, made that fall necessary. But why should not God embrace in his plan that great event, the fall of man, which he foresaw in the future? Would it have been wiser and better to leave out of account that most stupendous fact in the history of the human race? This same objection, which Mr. Hume and many others have brought forward, lies with equal force against the great central fact of the gospel, the death of Christ. God's plan touching this world included the incarnation and death of his Son. Jesus, the 'Lamb of God,' is spoken of as 'slain from the foundation of the world.' Rev. xiii. 8. But the incarnation and death of Christ presuppose the apostasy of the human race. Did this plan touching Christ make the apostasy of man a necessity? If preparing a world—fallen, so to speak, beforehand—for a race which God foresaw would fall, be inconsistent with his righteousness, it must be equally inconsistent to prepare a Saviour beforehand for that same race.

"Again, the divine plan touching the death of his Son included his betrayal by Judas and his crucifixion by the Jews. If Judas had known that God had poised the salvation of man upon the pivot of his treachery, he would doubtless have argued as Mr. Hume and others are accustomed to do. But did God's plan excuse his treason against his Lord? His own conscience, piercing and rending his soul with remorse, drove him to self-destruction, and Christ confirmed the sentence of his conscience and called him the 'son of perdition.' The fact that God weaves the foreseen crimes of men into his plans is no palliation of their guilt.

"Would it be wise and well to take no account of foreseen events? Jesus has gone to prepare mansions for those who will, as he foresees, believe in him: why not make provision for foreseen evils also? Our civil government, knowing the liability to crime among men—a liability which the experience of man has shown to be a practical certainty—makes provision for those crimes by maintaining a police, reform schools, prisons, and armies. The Governor of the universe, knowing the liability of man to sin and fall—a liability which by his foreknowledge was to him a certainty—made provision for that foreseen apostasy. He made provision, both by the creation of a world suited to a sinful race kept under a probation of mercy, and by appointing a Redeemer, the 'Lamb of God,' slain, in the eternal purpose, before the foundation of the world. If Mr. Hume's objection has force at all, it has force against every wise provision of God to meet the consequences of man's foreseen wickedness. It is wise, forsooth, on man's part, to foresee coming evil and prepare for it; but if God do this, men count it worse than folly: they declare it to be an endorsement of the evil! So foolishly do men reason about the high things of God! My answer to Mr. Hume, then, has four parts:

"1. The existence of unexplainable difficulties does not disprove the truth and reality of any fact or principle.

"2. The supposition that God made provision for the present apostasy of the human race is burdened with fewer and smaller difficulties than its denial.

"3. The word of God declares that he did make provision for the fall of man by the pre-appointment of a Redeemer.

"4. That style of reasoning which seeks to justify or palliate man's first sin because God prepared this world for a fallen race would palliate and justify all wickedness, because the sins of men are woven into every figure of the web of divine providence. Not the treason of Judas alone, but the whole sum of man's evil-doing, is embraced in the far-reaching plan of God. How this magnifies the wisdom of God! He binds together in one bundle his own righteousness and the sins of men, in a most intricate interlacing, yet without blending the two and without staining the glory of his holiness.

"I hope I have made this plain. Do you think, Ansel, that you can repeat the substance of this answer to Mr. Hume?"

"I will try, sir, if he asks."

"You will all notice," added Mr. Wilton, "that I have not denied that there is a deep mystery in this preparation for the sins of men not yet created, and that I have not attempted to explain this mystery. I have only tried to show that the admission of the view I have given you is more satisfactory to reason than its denial, and that the mysteries of this view are not unreasonable and self-contradictory, for the greatest mysteries are often the most reasonable things in the world.

"My introduction has become much longer than I designed, but now let us turn our attention to the subject of the lesson.

"To aid us in understanding God's wise arrangements in the management of heat, we need, first, to consider what heat is and to review the laws of its action. Without this, we could look on and wonder at God's working in nature, but could not explain that which we saw.

"Ansel, will you state the theories which have been held touching the nature of heat?"

"I will do it as well as I can. The ancient philosophers supposed fire to be one of the four elements of which all bodies were composed. The three other elements were earth, air, and water. These four elements were mingled in various proportions. Of these, fire was esteemed the purest and most ethereal; this constantly tended upward to the empyrean, the highest heaven, where the element of fire and light was supposed to exist unmingled and pure. In the seventeenth century, Beccher and Stahl, two German chemists, brought forward what is known as the *phlogistic hypothesis*.

They supposed that every combustible body held in composition a pure, ethereal substance which they called *phlogiston*, a Greek word which signifies *burned*, and that in combustion this phlogiston escaped. Flame was supposed to be this escaping phlogiston. These were the notions held about fire and combustion, but they are hardly worthy to be called theories of heat. The discovery of oxygen by Dr. Priestley of England, in 1774, and the introduction of the balance by Lavoisier of France, joined with the ever-enlarging circle of facts to be explained, rendered the phlogistic hypothesis untenable, and it was thrown aside.

"Until a few years since the *caloric* theory was generally received. According to this theory, heat is a *substance*, a subtle ether, diffused through all bodies and surrounding their atoms. This ether has been supposed to have a strong attraction for the atoms of every other substance, while between its own atoms a strong repulsion exists. In solid bodies each atom of matter, or in compound bodies each cluster of atoms, has been supposed to be surrounded by a little atmosphere, so to speak, of caloric, which prevented the atoms from coming into absolute contact.

According to this theory, heat expands bodies by increasing and deepening these minute atmospheres, thus pressing the atoms farther from each other."

"You need not explain this theory farther," said Mr. Wilton; "we have hardly time to go into the history of theories. Tell us the latest received theory."

"The theory now commonly believed is called the *mechanical* or *dynamic* theory. According to this theory, the essence of heat is *motion*. A hot body is one whose atoms are in a state of rapid and intense motion or vibration; and the sensation of heat on touching a hot body arises from the impact, or rapid blows, of the agitated atoms, communicating the same atomic vibration to the flesh and nerves of the hand."

"Very well stated, Ansel. This is the theory now more commonly received. The caloric theory, like the crude notions of the old Greek philosophers about fire, and like the phlogistic hypothesis, has been rejected because it failed to explain the phenomena of heat. Whether the dynamic theory is destined to share the same fate remains to be seen. It seems, however, to have a better foundation than its predecessors. The dynamic theory, though recently made popular, is by no means a recent conception. It was advocated by such men as Bacon, Newton, Rumford, Davy, Locke, and others. Locke, the distinguished intellectual philosopher who lived in the latter half of the seventeenth century (born 1632, died 1704), said, 'Heat is a very brisk agitation of the insensible parts of an object, which produces in us that sensation from which we denominate the object hot, so that what in our sensations is heat in the object is nothing but motion.' Benjamin Thompson, an American gentleman who went to Europe in the time of our revolution, and for his scientific fame was made Count Rumford, and became the founder of the Royal Institution of England, declared that he could form no conception of the nature of heat generated by friction unless it were motion.

"A beautiful generalization has been made to show how well this idea of heat harmonizes with the entire plan of the universe. In the whole boundless universe each system of worlds, like our solar system, may be regarded as a molecule, or complex atom. These cosmical molecules, or complex atoms of the universe, are in motion through unmeasured space. In these systems of worlds the planets, with their satellites, are the molecules, and they are in motion—indeed, they commonly have several motions. Our earth, for example, rotates upon its axis once each day; it revolves in its orbit around the sun once each year, and the axis of the earth has a slow wabbling motion which produces the precession of the equinoxes, requiring 25,868 years for a complete revolution. The earth also is made up of parts, and all these are in ceaseless motion. As said the old Greek philosopher, 'All things flow'—that is, everything is in a state of change. Solomon has well described this perpetual movement and change: 'One generation passeth away, and another generation cometh. The sun also ariseth, and the sun goeth down, and hasteth to his place whence he arose. The wind goeth toward the south, and turneth about unto the north. It whirleth about continually, and the wind returneth according to his circuits. All the rivers run into the sea; yet the sea is not full; unto the place from whence the rivers come, thither do they return again. All things are full of labor; man cannot utter it. The eye is not satisfied with seeing, nor the ear filled with hearing. The thing that hath been, it is that which shall be; and that which is done is that which shall be done, and there is no new thing under the sun.' Eccles. i. 4–9. It is certainly in harmony with this universal movement that the atoms of matter, though they seem so closely packed, should in their inconceivable smallness through inconceivably minute spaces vibrate, or rotate, or revolve through an orbit, never at rest. Intensity of heat we may think of as intensity of this atomic motion—a wider swing, so to speak, in their vibration or revolution. This, of course, requires a wider separation of the atoms and a consequent expansion of bodies. A feebler atomic motion permits the atoms to approach each other. In this manner we explain the enlargement of bodies by heat and their contraction by decrease of temperature. 'The ideas of the best-informed philosophers are as yet uncertain regarding the exact nature of the motion of heat, but the great point at present is to regard it as a motion of some kind, leaving its more precise character to be dealt with in future investigation.' This is the most we can do at present."

"What is the evidence," asked Samuel, "that the dynamic theory of heat is true?"

"The evidence that any theory is true is its ability to explain the facts or phenomena with which it has to do. If it explains all the facts and contradicts no known principles, it is regarded as true, or at least no objection can be made to it. Let me illustrate. Astronomers had long inquired what force or law controlled the movements of the heavenly bodies. At length Newton answered, A force of attraction between bodies which decreases in proportion as the square of the distance between them increases. This explanation has been found sufficient to explain all the known facts in the working of the heavenly bodies. Upon the basis of this theory astronomers calculate the positions of planets and comets for years and centuries to come.

"This theory led to the discovery of the planet Neptune, the last discovered of the primary planets. For thirty years irregularities in the motion of Uranus had been noticed. These variations were so slight that if another planet had revolved in the proper orbit of Uranus they would have seemed to the naked eye, throughout their course, one and the same star.

This slight irregularity of motion was so nicely measured that the place of the unseen planet which caused it was almost exactly calculated from the estimated force and direction of its attraction. This theory of a universal attraction of gravitation so well explains all the facts in the case, and has become so universally received, that we are liable to forget that, after all, it is nothing but a theory.

"Our idea of the structure of the solar system was at first only a theory. The astronomer does not see the planets revolving in regular circles through the heavens and moving around the sun. He only sees the shining points moving back and forth upon the concave vault, doubling and crossing their tracks apparently in the greatest disorder. How shall their motions be explained? Astronomers have found that the motions of planets revolving around a central sun, when seen from one of the planets, must present just these apparent irregularities. This explanation is so full and complete that it is now counted not a theory, but an established fact. The same may be said of the shape of the earth.

"The dynamic theory of heat explains the

phenomena of heat better than any other explanation that has been proposed. It explains the radiation of heat from the sun or from any other hot body: vibrations or impulses are propagated through that ether which is supposed to fill all space. It explains the conduction of heat through solid bodies in the same manner. It explains the expansion of bodies: the atomic motion forces the atoms of bodies farther apart. It explains the production of heat by friction or collision, which no other theory is able to do: the shock of the collision generates this atomic vibration. It explains the production of heat by combustion: the atoms of oxygen and carbon or hydrogen dash against each other and generate heat by the collision. This theory explains the transmutation of motion, or living force, and electricity, into heat, and the transmutation of heat into electric or mechanical force. These points will come up again, and I now only refer to them in answering Samuel's question. The dynamic theory explains the phenomena of heat and its relations to force, light, and electricity exceedingly well, and for this reason men look upon it with favor and count it as probably true. If in the progress of scientific investigation it shall be found to explain all the new facts discovered and meet well all the demands made upon it, it will at length be received as an admitted principle in physical science. The *wave* theory of light and the *vibratory* theory of sound may be looked upon as thus established.

"At our next lesson we shall take a rapid review of the effects and laws of heat."

CHAPTER IV.

HEAT A GIFT OF GOD.

THE class is again promptly in place and ready for work.

"As I announced a week ago," said Mr. Wilton, "we will to-day take a rapid review of the effects and laws of heat. Will you tell us, Peter, the first and chief of these effects?"

"Yes, sir: combustion."

"What is combustion?"

"Commonly the rapid union of oxygen with some combustible substance, attended with the evolution of heat."

"Was your answer correct, then?"

"No, sir," said Peter, blushing; "I spoke before I thought."

"Will you correct your answer?"

"The first and chief effect of heat is expansion."

"That is right. Our sensation of heat is of

course only a *sensation*—merely the *feeling* which results from the effects of heat upon our nerves—but the chief physical effect of heat is the expansion of bodies. The chemical qualities of bodies are not changed: they are not made either heavier or lighter. A sufficiently high temperature renders bodies luminous, and then we call them red hot or white hot. Solid bodies begin to be luminous at a temperature of about one thousand degrees. But the one invariable effect of heat, with two or three apparent exceptions, is expansion. You may mention, Samuel, some familiar illustrations of the effect of heat in expanding bodies."

"The blacksmith heats the wagon-tire in order that it may easily slip over the wheel. If a kettle be filled with cold water, by heating it the water is expanded and runs over. I have noticed that the spaces between the ends of the successive iron rails upon the railroad are larger in winter than in summer, showing that the rails are shorter in winter than in summer. While skating during the cold winter evenings upon the mill-pond, I have seen cracks in the thick ice start and run across the mill-pond with a roar almost like thunder. The ice was contracted by the cold till it could no longer fill the whole space between the banks, and being frozen fast to the banks, it was torn asunder. The mercury in the tube of a thermometer is constantly expanding or contracting by every change of temperature."

"Yes, those are all good illustrations, and we might go on to mention others equally good by the score. In cold countries, during the intense cold of winter, the surface of the earth cracks by shrinkage, just as you have seen the ice upon the mill-pond torn in two. The Britannia iron tubular bridge over the St. Lawrence at Montreal rises and falls two and one-half inches on account of greater expansion of the upper surface when exposed to the heat of the sun, while a loaded freight train causes a depression of but one-fourth of an inch. A few years since, in order to make some philosophical experiments connected with the rotation of the earth upon its axis, a ball was suspended by a wire in the interior of Bunker Hill monument. By this means it was accidentally discovered that the heat of the sun, expanding the sides of the monument exposed to its rays, caused the whole monument to sway back and forth daily."

Here Ansel raised his hand.

"What is it, Ansel?"

"I was going to mention the belief of geologists that the mountain ranges were thrown up by the contracting of the earth's crust on account of cooling."

"That is an illustration of contraction by loss of heat on an enormous scale. The materials which form our globe may have existed in the beginning in a nebulous or gaseous state. There is certainly very good reason for believing that the earth was once in a fluid state, the whole of its substance molten by intense heat. It is certain that the interior is now hot, and portions of it molten. It is by very many believed that the whole interior is molten. The crust of the earth may have been formed by cooling. If after an outer crust had been formed, and its temperature had fallen so low as to become nearly stationary, the interior mass continued to cool, the molten mass would tend to sink away from the crust and the crust would sink in upon it by wrinkling. Thus mountains may have been formed. Along the line of fracture the easiest vents would be formed for volcanoes. But this carries us somewhat aside from our subject, and as the expansion of bodies by heat has been sufficiently illustrated, we will leave it. Will some one now state the manner in which the dynamic theory of heat explains this expansion?"

Samuel answered: "I think you have already given us the explanation."

"I have briefly referred to it, but you may give it again."

"The atomic motion which is supposed to constitute what we call heat, whatever that motion be, whether a vibration or rotation or revolution, requires that the atoms of bodies shall not be packed in absolute contact, and the more intense the agitation or the wider the swing of the vibration or revolution, the greater must be their separation. Hence heat expands bodies by thrusting their atoms farther apart."

"That will do," said Mr. Wilton. "Let us look now at some of the secondary effects of heat. You may mention some of them, Ansel."

"Heat relaxes or overpowers the cohesive attraction of bodies."

"What is cohesive attraction?"

"It is that force which binds together the

atoms of matter in simple substances, that is, bodies like iron or copper or silver, composed of but one kind of substance, or in compound bodies it is the force which unites the compound molecules of matter."

"Give us now some illustrations of the effect of heat in overcoming cohesive attraction."

"The blacksmith heats his iron in order to overcome its cohesive attraction and render it soft, that he may easily hammer it. The founder heats his metal till its cohesion is so far destroyed that it becomes fluid and can be poured into the mould. Heat relaxes the cohesive force of ice and changes it to water, and by farther heating its cohesion is entirely overcome and the water is changed to a gas."

"We use heat also in cooking our food," spoke up Peter: "is it not because heat destroys the cohesive attraction, and thus softens it?"

"If that were the only effect of heat upon food," said Mr. Wilton, "we should be obliged to eat our food hot, for as soon as it cooled the cohesion would return and the food would be raw again. The operation of heat in cooking is various, and part of the effect is commonly to be ascribed to the water in which the food is cooked or to that which is contained in it. By the combined agency of heat and water starch swells to twenty or thirty times its original bulk and the minute starch grains burst open. In cooking potatoes the starch of the potato absorbs a portion of the water that is in it, and thus renders it dry and mealy. The action of heat and water upon rice, wheat, and other grains is similar to their operation upon starch. In the baking of bread the starch is converted into gum. In boiling flesh the effect is partly due to the solvent powers of water: the juices of the flesh are extracted, the gelatin is dissolved, the fat is liquefied, and the cells in which the fatty matter is held more or less burst, the albumen is solidified, and by long boiling the texture and fibre of the flesh are destroyed. The albumen of an egg, that is, the white, coagulates by heat. But in most of these processes the action of heat cannot be separated from that of water.

"But there is another effect of heat very important both in nature and in the arts. What is that?"

"The quickening of chemical affinity," answered Samuel.

"That is right: heat is necessary for the operation of chemical affinity. Perhaps this is only a weakening of the cohesive force, thus allowing the chemical attractions to assert their strength. But the fact is that, while in many cases the chemical affinities act with great energy at ordinary temperatures, in other cases they slumber, however closely the substances are brought into contact, till their temperature is raised. Samuel, you may mention some illustrations of this principle."

"A few months ago I visited Hazard's powder mills, in Enfield, Connecticut, and there learned how gunpowder is made. The charcoal, the sulphur, and the nitre are first finely pulverized, then ground together for hours till thoroughly mixed, and afterward pressed together. This mass is then broken into grains and the grains polished. But though these elements are brought into so close contact, yet they do not combine and explode till heat is applied. The same is true of the combustion of wood and coal. The carbon and the hydrogen of the fuel are constantly surrounded with the oxygen of the air, but they do not take fire and burn, that is, they do not combine with the oxygen, till they are raised to a red heat, or perhaps even to a higher temperature. If a stove filled with burning coal be cooled down to a low temperature by applying ice, the combustion will cease, the fire will go out. Our teacher at the academy on one occasion heated a steel watch-spring red hot and plunged it into a jar of oxygen, and the steel spring began quickly to burn with great fury."

"You have given us good illustrations, Samuel, and that which is true of carbon and hydrogen and oxygen is true of substances in general. The effect of heat in producing chemical changes is very important everywhere. It is seen not only in the chemist's laboratory and in the artisan's shop, but also in the laboratory of Nature. Plant a grain of corn in midwinter: why does it not germinate and grow? Nothing is needed but the requisite heat to quicken the chemical affinities into action. Earth and air furnish the needed material for the growth of forest trees in winter as well as in summer, but the cold holds in check the chemical forces and prevents the requisite chemical combinations. No sooner does the sun quicken that atomic vibration or revolution which we call heat than vegetable growth begins. Heat is necessary for those chemical changes by which food is digested in the stomach and the processes of nutrition carried on in every part of the body. If a man finish his dinner with ice cream or ice water, the process of digestion is delayed till the contents of the stomach recover their proper temperature. This is one chief reason why warm, comfortable clothing is so very important, especially for children. All the vital processes are chemical processes: they are carried on through chemical affinities. Unless the body be kept at a suitable temperature, these processes are feeble and imperfect, nutrition and vital combustion are hindered, and diseases are engendered.

"These, then, are the chief effects of heat. It expands bodies, weakens cohesive attraction, and quickens the chemical affinities into activity."

Ansel again raised his hand.

"What do you wish?"

"Will you please tell us, Mr. Wilton, how this weakening of cohesive attraction is explained upon the dynamic theory of heat?"

"I will do so with pleasure. The increased atomic motion in the heated body throws the atoms farther apart, as we have already learned, and by this increase of distance their attraction is diminished. If the earth were twice its present distance from the sun, their attraction for each other would be four times less than it now is; if its distance were three times as great, their attraction for each other would be nine times less. The attraction of gravitation diminishes in proportion as the square of the distance through which it must act increases. Perhaps cohesive attraction diminishes according to the same law, though the spaces are so small that this cannot be demonstrated, but it is certainly weakened by the expansion of bodies through the agency of heat."

Here Peter raised his hand.

"What will you say, Peter?"

"Do not men heat and burn bricks, not to soften them, but to harden them?"

"That is true," said Mr. Wilton; "but in this there is a process of drying as well as of heating, and the hardening is due chiefly to the complete drying by the intense heat. Too great heat will melt bricks while in the process of burning. I once heard a brick-burner say that he could melt the brick around the arches in his kiln in half an hour, if he pleased to put in fuel and let the fire burn. Indeed, almost every known solid substance has been fused by heat. Whether carbon has ever been melted is an unsettled question."

"I would like to inquire," said Samuel, "why water will not burn. Is it because it evaporates before it reaches a sufficiently high temperature?"

"This is a little aside from our subject, but the incombustibility of water is a provision of the Creator so very important that we will stop to notice it. I think, however, that by a little thought you yourself can answer the question. Tell me again what combustion is."

"Combustion is commonly the combining of oxygen with some other substance called a combustible. The rusting of iron and the decay of organic bodies are forms of slow combustion."

"Now tell us the composition of water."

"Water is composed of oxygen and hydrogen —eight parts of oxygen to one of hydrogen, by weight, or two parts of hydrogen to one of oxygen, by measure."

"How is water formed from these two gases? Are they mixed together as oxygen and nitrogen are mingled in the air, or are they chemically united?"

"They are chemically united: they are burned together. When hydrogen burns, the product is water."

"Water is then a *product* of *combustion*. Can you not now tell why water is incombustible?"

"I think I now see the reason. The oxygen, being itself the supporter of combustion, will not burn, and the hydrogen has been already once burned in the formation of water."

"And that which is true of water is true, in a greater or less degree, of other products of combustion. The burning of charcoal produces carbonic acid, and carbonic acid will not burn because it is the production of combustion. A candle is extinguished by it as quickly as by water. By a recent invention carbonic acid is used to extinguish conflagrations. The carbon has once united with oxygen, and a second combination with an additional amount, or, as a chemist would say, with another equivalent, of oxygen is much more difficult."

"I think," said Samuel, "I now understand why water will not burn, but will you please also to tell us why water puts out fire better than almost anything else?"

"In order to extinguish fire one of two things must be done: either the supply of oxygen must be cut off or the combustible must be cooled down to a temperature below the burning point, when the combustion will cease of itself. When we shut the draught of an air-tight stove, we check the combustion by shutting off the full supply of oxygen. If we could wholly prevent the access of oxygen to the fuel, the fire would at once be extinguished. If oxygen should then be admitted again before the fuel had cooled down below the burning point, combustion would at once begin again. A blazing brand is extinguished by being thrust into ashes, because it is shut away from oxygen. In the same way we extinguish the flame of a candle with a tin extinguisher. On the other hand, fires often go out because the necessary temperature is not maintained. Water puts out fire in both these ways, but especially by the second. Water poured in torrents from a fire engine upon a fire forms a film of water, and the burning material shuts out the oxygen. But the water acts chiefly by lowering the temperature. No other known substance except hydrogen gas requires so much heat to raise it through a given number of degrees of temperature as water. As much heat is required to heat one pound of water as thirty pounds of mercury. Hence, water poured upon burning timber cools it to so low a temperature that it ceases to burn.

"In addition to this, we may notice that wood saturated with water cannot be heated above the boiling point of water till the water is evaporated. As fast as the wood and the water rise or tend to rise above two hundred and twelve degrees, the water changes into steam and carries away the additional heat. The consumption of heat in the formation of vapor we must look at more carefully in a future lesson. We will suppose that a house is in flames. A fire engine throws a stream of cold water into the midst of the conflagration. The cold water, dashing against the burning wood, cools the heated surface; it is absorbed into the pores of the wood and hinders its rapid heating; a portion of the water, being changed into steam, carries off the heat; the steam, mingling with the flame, lowers the temperature of the burning gas, and in proportion as steam fills the surrounding space oxygen is driven away. A burning coal mine in England was once extinguished by forcing steam into it, thus driving out the air which supported the combustion and cooling down the burning coal.

"The advantages which men receive from these agencies of heat are so manifest that we cannot help noticing them. I do not refer to the comfort of a pleasant temperature, nor the impossibility of living in a temperature extremely low, but to all those processes by which man subdues nature, provides for himself food, clothing, and dwelling-places, and builds up civilization. Heat is that force which enables man to accomplish his ends. Heat brings the iron from the native ore, and heat renders it malleable and plastic to be shaped for man's uses. Heat quickens the chemical affinities and renders the arts of civilized life a possibility. Heat brings together oxygen and carbon in ten thousand furnaces, and the heat engendered by the combustion, changed to force, drives the ponderous or nimble machinery which carries on the work of the world. Heat quickens the chemical affinities and causes the wheat to grow; heat prepares the wheat for man's food;

and by the aid of heat that food is changed in man's body, nutrition goes on, the body is built up, waste matter is removed, and all the vital processes are supported. Without these agencies of heat—softening and subduing stubborn matter on the one side, and quickening its forces on the other—man could not exist.

"Let me remind you that these agencies of heat are of God's devising. If the operations of heat are beneficent to man, it is because God wished to bless his creatures. I am not much given to moralizing, but when I see how completely these simple effects of heat meet man's wants, I cannot help remembering and admiring the wisdom of the great Designer. It is *God* and not blind, unconscious Nature that is working."

"This reminds me," said Samuel, "of the tradition in Greek mythology that Prometheus stole fire from Jupiter and brought it down to man in a reed as a precious treasure. It seems to me like a gift from heaven."

"This mythological tradition has, however, one falsehood: there was no need that men should steal fire from the gods; God freely gave it. Heat is indeed a gift from heaven."

CHAPTER V.

CONVEYANCE AND VARIETIES OF HEAT.

"TO-DAY we review the modes in which heat passes or is conveyed from place to place. It is evident that if heat were confined to the very place or point where it is generated, it could subserve none of those uses to which it is now applied in the economy of Nature or in the works and arts of man. But heat passes from place to place with great facility, and by one method, with the speed of light, it tends to diffuse itself evenly through all; it seeks an equilibrium. The modes of its diffusion, or conveyance, are three in number. Ansel may name them."

"Heat passes from place to place and from body to body by 'conduction,' by 'radiation,' and by 'convection.'"

"What is meant, Ansel, by the 'conduction' of heat?"

"The passing of heat from atom to atom and from particle to particle through a body is called conduction."

"That is right. I will call upon Peter to give some illustrations of the conduction of heat."

"The examples are so many," Peter answered, "that I hardly know what to mention first. If I hold a pin in the flame of a lamp, the part of the pin that touches the flame is first heated, but soon the heat runs along the whole length of the pin and burns my fingers. The parts of a stove which touch the fire are first heated, and from them the heat spreads through the whole stove. A pine-wood shaving, kindled at one end, is heated by conduction, but the heat passes through it very little faster than the flame follows. Heat escapes from our bodies by being slowly conducted through our clothing. There is no end to the examples of conduction which one might give."

"We must not think of the conduction of heat," said Mr. Wilton, "as if it were a fluid slowly absorbed by a porous body, as water poured upon the ground soaks into it, or as water percolates through a lump of sugar and moistens the whole of it. We must remember that the transfer of heat is not a transfer of any substance, but a transfer of motion. One atom is set in motion, and strikes against another atom and sets that in motion, and thus motion is communicated from atom to atom and from molecule to molecule through the whole mass of matter till every atom is agitated with the heat vibrations. Do all bodies conduct heat with equal rapidity?"

"No, sir," replied Ansel; "there is the greatest possible difference. Some substances are called good conductors, because heat permeates them so readily and rapidly; others conduct heat very slowly, and are called poor conductors or bad conductors."

"That is right. Every child soon learns by experience to make a practical distinction of this kind. He very soon understands that he can hold a stick of wood without burning his hand, even though it be blazing at the other end, but that when a piece of iron is red hot at one end he must not take hold of it at the other. The child very soon learns to know the different feeling of a cotton night-gown from one of flannel, and the difference in apparent warmth between a linen pillow-case and a woolen blanket.

After a room has been heated for a considerable time the various objects in it all become of the same temperature, and the same is true in a cold room; but how great the difference in the sensations produced by touching the oil-cloth and a woolen carpet in a cold room! Good conductors of heat, if hot, feel very hot; or if cold, feel very cold; while poor conductors make a much less decided impression. Why is this, Samuel?"

"The good conductors receive heat or part with it very readily. If the good conductor be hotter than our bodies, it imparts its heat rapidly to our hand, and because we receive heat rapidly from it, it feels to us very hot. Or if it be colder than our bodies, it takes heat from our hands very rapidly, and gives the impression of being very cold. Poor conductors impart heat to the skin or take it away more slowly, and hence feel as if their temperature were more nearly like that of the body."

"The conducting qualities of bodies," said Mr. Wilton, "seem to depend chiefly upon their structure or the arrangement of their atoms. Bodies which are compact and solid in their structure convey heat more rapidly than those which are loose and porous. Hence solids are better conductors than fluids, and fluids are better conductors than gases, and among solids the metals are better conductors than organized bodies, like wood or flesh, and better than the loose and porous minerals. In bodies of loose, porous, or fibrous texture, the continuity of the conductory substance is constantly broken. The particles in a mass of sawdust touch only at a few points, leaving frequent spaces. In woolen and cotton fabrics the points of junction of the fibres are very few, comparatively. For this reason the motion is not readily communicated from atom to atom.

"The crystalline arrangement of atoms has an influence upon conduction of heat. Heat is conducted more rapidly in a direction parallel with the axis of crystallization than across that axis. Wood conducts heat more rapidly in the direction of the grain. This arrangement seems to be well adapted for keeping trees warm in winter. Their roots reach down into the earth, which remains warm in the coldest weather. This heat of the earth travels along the fibres up through the tree, while the heat conducted across the fibres escapes much more slowly into the open air.

The bark also, being a very bad conductor, hinders the escape of heat. Of metals, silver is the best conductor. I will give you a brief table which will show the great difference in the conducting qualities of some of the metals. Counting the conducting qualities of silver as 100, the table is: 'Silver, 100; Gold, 53; Copper, 74; Iron, 12; Platinum, 8; German Silver, 6; Bismuth, 2.'—*Youmans*.

"What is the second method by which heat passes from place to place?"

"It is radiated," replied Ansel.

"And what is radiation?"

"It is motion in straight lines or rays diverging from a centre. From a hot body heat is passing off in straight lines in every direction. As a lamp radiates light, so does a hot body radiate heat."

"Radiant heat," said Mr. Wilton, "moves with the same velocity as light, that is, one hundred and ninety-two thousand miles per second. It also follows the same general principles as light in all its motions. It is absorbed, reflected, or transmitted in the same manner as light. And this is true of either luminous heat—that is, heat radiated from a body which is red hot—or obscure, or dark heat.

"As there are good and poor conductors, so there are good and bad radiators of heat. The radiation of heat depends upon three conditions:

"1. Upon the temperature of the body. The higher the temperature, the more rapid and energetic is its radiation.

"2. Upon the surface of the radiating body. A dull, rough surface radiates heat more rapidly than a surface bright and polished.

"3. Upon the substance of the radiating surface. With surfaces equally smooth and bright, some substances radiate heat much better than others. A surface of varnish radiates heat much more powerfully than a surface of gold or silver.

"Ansel, you may, if you can, explain the radiation of heat."

"I can give no other explanation than that radiation is conduction through that subtle ether which is supposed to pervade all space."

"Very well; perhaps that is as good an explanation as can be given. But it seems rather like the propagation of an impulse than the spreading of atomic vibrations in every direction. The motion is propagated in straight lines. If it be conduction, it must be carried on by different vibrations from those of ponderable substances. Heat, light, and electricity are supposed to be all propagated through the same theoretical ether. Sir Isaac Newton estimated the density of the ether as seventy thousand times less than the density of our atmosphere, and its elasticity in proportion to its density as four hundred and ninety millions times greater. But the very existence of this universally-diffused ether is a supposition made to account for the phenomena of light, heat, and electricity; and, of course, all its qualities must be theoretical also. Radiation is believed to be the propagation of a motion or impulse through an inconceivably rare and elastic ether.

"Peter, what is the third method by which heat passes from place to place?"

"Convection," was his reply.

"What is meant by convection of heat?"

"The conveyance of heat by carrying a heated body. If I remove a hot iron or a kettle of hot water, I must of course carry the heat which it contains."

"A very good illustration of the convection

of heat," said Mr. Wilton, "is seen in the common method of heating water. The heat is applied at the bottom of the vessel containing the water; as fast as the water at the bottom next the fire is heated, it rises and carries the heat to the top; cold water comes to take its place, and this in turn is heated and rises and carries heat to the top. This process is carried on till all the water comes to the same temperature. Thus water is heated by convection of heat.

"A grander illustration is seen in winds and ocean currents. Warm winds carry heat enough to warm a continent, and the mighty ocean currents are still more efficient in transferring heat from one part of the earth to another.

"Another point we need to understand. When radiant heat falls upon a body, what becomes of it?"

"It is disposed of," answered Samuel, "in one of three ways: it may be reflected according to the same principles by which light is reflected; or it may be transmitted, that is, pass through the body; or it may be absorbed, that is, stop in it."

"Very well stated, Samuel. In regard to reflection I need to say very little. You know how light is reflected from a polished surface, such as a lamp reflector: heat is reflected in the same manner. One fact you must bear in mind touching reflected heat: it does not heat the reflecting body.

"There is no need of telling you that light passes through certain substances. It passes through gases and through some liquids and some solids. The best of glass, though it is so solid, interposes very little hindrance to the passage of light. Heat in like manner radiates through certain solids. Luminous heat is radiated through glass. Rock-salt transmits dark heat also. A plate of alum permits light to pass, but stops both luminous heat and dark heat. Remember that transmitted heat, as was said of reflected heat, does not heat the body through which it passes. I have seen boys make burning-glasses of ice. The heat passes through them and burns that upon which it is concentrated, while the ice itself through which the heat passes is not melted.

"If a body have a good radiating surface, that is, if its surface be dull and rough, the heat which falls upon it will be mostly absorbed. The reflecting and absorbing qualities hold an inverse ratio to each other; the better the reflecting qualities, the worse the absorbing, and the worse the reflecting, the better the absorbing. Heat which is absorbed by a body commonly raises its temperature, and remains in the body till it is slowly radiated or is conducted away by the air or other bodies which come in contact with it.

"What is that heat called, Ansel, which is absorbed by a body with no rise of temperature?"

"It is called *latent* heat."

"That is the old and common expression, but what is meant by latent heat?"

"The word *latent* signifies *lying hidden* or *concealed*. Latent heat, as you suggested in your first question, is that heat which a body receives without showing it by a change of temperature."

"That name 'latent heat,'" said Mr. Wilton, "expresses the opinion of those who invented it; they supposed that heat was in some manner hidden in certain bodies. We must not suppose, however, that this latent heat continues to exist in bodies as heat; latent heat is that heat which is converted into force or some other motion than the atomic heat vibrations, and is employed otherwise than in raising the temperature. You will understand this best by an illustration.

"Take one hundred pounds of ice at the temperature of thirty-two degrees, that is, as warm as is possible without melting. That one hundred pounds of ice will absorb heat which would raise one hundred pounds of ice water through one hundred and forty degrees, and by receiving that heat it is melted, but the water produced has the temperature of thirty-two degrees. It has received one hundred and forty degrees of heat, but its temperature is not raised a single degree. This one hundred and forty degrees of heat has been transmuted into force and employed in overcoming the crystalline attraction of the atoms of water.

"Let that ice water at thirty-two degrees of temperature receive one hundred and eighty degrees of heat, and the water rises to two hundred and twelve degrees, the temperature of boiling. But whatever additional heat is absorbed brings no increase of temperature, but transforms the water into steam. It is employed in overcoming the cohesive attraction of the molecules of water and changing the liquid to a gas. About one thousand degrees of heat is thus expended, but the steam which is produced has only the temperature of two hundred and twelve degrees. If the process be reversed, the steam gives up, as it is said, the one thousand degrees of heat in returning to the condition of water and the one hundred and forty degrees in resuming the crystalline structure of ice. The heat which was employed as force in overcoming the atomic and molecular attractions is transmuted again to heat, and shows itself in raising the temperature. And that which is true of water is true of any other substance in changing its form from a solid to a liquid or from a liquid to a gas, or the opposite. In an amount different for each kind of matter, in all these changes of condition, heat is transmuted to force or force to heat.

"These transmutations are going on ceaselessly in the operations of Nature, and without understanding them we cannot appreciate the wonderful operations of heat in the world. The heat of the sun beams upon the ocean; the greater part of that heat is expended as force in overcoming the molecular attraction of water, thus converting it to vapor, and in raising that vapor to the higher regions of the atmosphere. This heat-force, or, as we might call it, 'sun-

Curiosities of Heat.

power,' expended upon the earth, amounts to thousands of millions of horse-power daily.

"Examples of the transmutation of force into heat abound everywhere. A boy strikes his heel upon the stone pavement; from the point of contact between the stone and the steel points in his boot heel sparks of fire fly out. Force is changed to heat so intense that particles of steel are set on fire. Savages who have no better methods of kindling fire rub dry wood together till the sticks ignite. The force expended in overcoming the friction is changed to heat. In the combustion of coal beneath the steam boiler we see both processes going on. The atoms of carbon dash against the atoms of oxygen, and the force of the collision generates the heat of the combustion. This heat, born thus of force, is again transmuted to force, and drives the engine and the machinery attached. In our study of God's management of heat we shall constantly meet with these changes. You will need, therefore, to study carefully this subject of latent heat.

"Dr. Joule, of Manchester, England, has discovered the ratio between heat and force, that is, the amount of force which by transmutation produces any given amount of heat. The force of a one-pound weight which has fallen one foot is taken as the unit of force, and the amount of heat which is required to raise one pound of water one degree is taken as the unit of heat. By many and various careful experiments, Dr. Joule demonstrated that 772 units of force are the equivalent of one unit of heat. A pound weight falling 772 feet, or 772 pounds falling one foot, and then arrested, produces heat sufficient to raise one pound of water one degree. The result is the same whatever the method by which the force is expended. If water be agitated or shaken, if sticks of wood or iron plates be rubbed together, if an anvil be struck with a hammer, or if a bar of iron or copper be moved back and forth between the poles of an electromagnet, the force expended is changed to heat. You must remember, however, that force becomes heat only so far as the force is actually expended, or used up so that it no longer exists as force.

"These conclusions are supported by other beautiful experiments. 'An electric current which, by resistance in passing through an imperfect conductor, produces heat sufficient to raise one pound of water one degree, sets free an amount of hydrogen which, when burned, raises exactly one pound of water one degree. Again, the same amount of electricity will produce an attractive magnetic force by which a weight of 772 pounds may be raised one foot high.'—*Youmans.* We conclude from experiments like these that heat, mechanical force, and electricity are interchangeable forces; they may be transmuted the one into another.

"By this principle of the transmutation of heat and mechanical force we explain the production of heat by compression and the loss of heat by expansion. Samuel, you may state the fact upon this point."

"If any substance be suddenly compressed," answered Samuel, "heat appears; if it be expanded, cold is produced. Since gases expand or yield to pressure so readily, they furnish the best illustration of this principle."

"The suddenness of the compression or expansion," said Mr. Wilton, "is a matter of no consequence. The effect is the same whether the operation be sudden or slow, but if the compression or expansion be slow, the heat or cold generated is less apparent; the heat is dissipated as fast as produced and the colder gas is warmed by the vessel which contains it. Ansel, how shall we explain this?"

"I cannot explain it, sir."

"The explanation is very simple," said Mr. Wilton. "Mechanical force is employed in the compression of the gas; the force is expended and used up upon the gas, and appears again in the form of atomic heat motion. In the expansion of gases the operation is just the reverse; the atomic heat motion is expended in producing expansion, and hence disappears as heat. The general principle is that no force can be expended in two ways at the same time.

"One other point we must notice to-day, that is, *specific heat*. What is understood, Ansel, by this term, specific heat?"

"The relative amount of heat which different substances require to raise their temperature through any given number of degrees."

"That is right. I think that you all must have noticed that it requires much more heat to raise the temperature of some bodies than others. What an amount of heat is required to raise the temperature of water! That heat which will raise one pound of water one degree will cause an equal increase of temperature in five pounds of sulphur, or four pounds of air, or nine pounds of iron, or eleven pounds of copper, or thirty pounds of mercury, lead, or gold. This is what is meant by saying that one substance has a greater capacity for heat than another. The specific heat of water is greater than that of any other known substance except hydrogen gas. This fact, taken in connection with its great specific latent heat and its poor conducting qualities, renders it exceedingly important in regulating climate and moderating extremes of temperature; of this you will be reminded very often as our lessons go on.

"No law or principle determining the specific heat of the various elements and explaining the different capacities for heat has as yet been discovered. It has been suggested that specific heat depends upon the number of atoms, that it holds an inverse ratio to their combining numbers, or, what is the same thing, a direct ratio to the number of atoms. This would harmonize well with the dynamic theory of heat, but the harmony between the specific heat of substances and the number of atoms is not sufficiently uniform to establish this supposition.

"This completes our review of first principles. I hope that this not very entertaining review of your academic studies has not wearied you of the very word *heat* and worn out your interest in examining God's management of heat before making a beginning."

"I think," said Samuel, "that we are not in the habit of becoming disgusted with our studies."

"You may expect," continued Mr. Wilton, "if the past has been interesting to you, that the lessons to come will prove more interesting still. Next week we shall consider the abundant provision which the Creator has made for warming the earth."

And let me say to you, patient reader, that if I had known that you were as familiar with the laws and principles of heat as Ansel, Peter, and Samuel seem to have been, this and the preceeding chapter would not have been written. However dull this review may have seemed to you, it was needful, perhaps, for others, that they might understand the wonderful works of God which we shall now proceed to examine. And, reader, do not forget that heat itself, that subtle motion and mighty force, with all its laws and principles, is one of God's works. Already have we been looking at the Creator's handiwork. Already have we been trying to trace out the thoughts of God as they are written in the "Bible of Nature." The thoughts of God are great and wonderful. It has been useful and interesting to read thus far in this book written with the finger of the Creator of worlds and of man, even if we turn not another page.

CHAPTER VI.

MANAGEMENT AND SOURCES OF HEAT.

WHILE the lessons which have been reported were going on, the religious interest in the church was deepening. Mr. Wilton did not cease to make his sermons instructive, but, in addition to the instruction, he made them more and more pungent and persuasive. He aimed to gather up the impressions and convictions already wrought in the minds of his hearers and combine them for united and immediate effect. He believed that this was to be a reaping-time.

Mr. Hume was becoming interested, not because he had been at church, for he had not been there, but the Holy Spirit of God was working upon his heart. He was becoming uneasy in his unbelief. For some reason, he knew not why, his opinions were becoming more and more unsettled. He did not like to go to the house of God; his self-will and pride of consistency rebelled against the thought of hearing and believing the gospel; but he was restless and discontented away from the place of worship. His associations with his infidel comrades grew distasteful. His Sundays were days of distress: with his attention relieved from business cares, thoughts of God and eternity pressed upon him, and he could not escape them. At length he determined to go and hear Mr. Wilton again: perhaps he should hear something which he could so positively reject as to set his mind at rest. He went, accordingly, the next Lord's Day, and heard a very impressive sermon.

The text for the forenoon was Ps. lxvi. 5: "Come and see the works of God: he is terrible in his doing toward the children of men." The sermon gave first a brief and rapid review of some striking displays of God's displeasure at the sins of men: that ancient world of men whose "thoughts were only evil continually" he overwhelmed with the flood; he burned with fire from heaven Sodom and Gomorrah, Zeboim and Admah, those lascivious and festering cities of the plain; he sent his torturing and consuming plagues upon the Egyptians, and sunk the army of Pharoah like a stone in the deep waters of the Red Sea: "they sank as lead in the mighty waters;" he caused the earth to open and receive Korah and his adherents, and bade his angel in "one night" to touch with death the thousands of Sennacherib's army. This record of divine wrath against evil-doers has startled the consciences of wicked men, and will continue to startle them so long as the ungodly live upon the earth. It is easy for unbelievers to call the word of God a record of fabulous wonders, but that record lives and will live, and its words assert their divinity by touching and burning the consciences of men as if they were tongues of fire.

"But to the thoughtful man," said Mr. Wilton, "there is a manifestation of God's displeasure at sin even more impressive than these miraculous judgments. The Creator has built his wrath against sin into the very fabric of the universe; he has written it upon the very atoms and elements of matter and of mind, and graved it upon the 'nature of things.' The forces of Nature are all instinct with holy wrath against ungodliness. Evil doing works out evil consequences by the regular course of nature. Babylon, Nineveh, and Tyre were great and prosperous, and as mighty in wickedness as in commerce and war. In the height of their prosperity God denounced upon them disaster and desolation, and by the natural processes of evil their decay and destruction came upon them. No miracle broke the harmony of their mighty march to decay and the silence of death. Great nations have perished, but not till they became corrupt. Rome fell, but luxury first gendered luxuriant vices, and vices enervated her hardihood and undermined the defences of her courage. No righteous nation ever perished. No nation ever fell into decay till ripe in sin and ready for moral putrefaction. But against wicked and corrupt nations wars and desolations are determined, and the end thereof is with a flood. The very forces of Nature seem allied in firm compact with the laws of God, ready with resistless hand to avenge their transgression and to visit evil upon evil-doers. This steady march of all the forces of the world in bringing decay and wretchedness upon sinners is more impressive than any single desultory example of avenging wrath.

"But perhaps an unbeliever replies, 'Not so; there is a natural law of development, decay, and death, apart from sin. Trees grow up, become old, and die. Men pass from childhood up to manhood, and from manhood down to second childhood, and return to the dust whence they came. By a like principle, nations pass through similar changes of development, decay, and desolation. But in all this there is no manifestation of divine favor or disfavor.'

"This is narrow and false reasoning. If a single great city had become corrupt while all the world beside remained righteous, and God had denounced his displeasure upon it and had executed his wrath by sudden and tremendous judgment, that one city standing out in single and solitary ungodliness and desolation, who would deny, who could deny, that the fate of that unhappy city was a manifestation of divine displeasure? If a second example were made of a second ungodly city, would the expression of divine wrath be weakened? Nay; every man would say that it is made stronger. What if a third example be made of a third city? What if every wicked city is made an example? What if God embody his displeasure at evil-doing in the structure of the world, and give to the very atoms of matter and the elements of mind such natures that by the working of their own proper forces, without a miracle, they shall bring pain and evil, decay and death, upon the ungodly? What is this but writing his wrath against sin upon the earth and sky, upon matter and the consciences of men, declaring by this that till the heavens and the earth and the spirits of men be no more he will never withdraw his indignation? This is what God has done. The wicked man sets in motion the machinery which works out his own everlasting undoing. His own hand sows the seeds of death, and as those seeds germinate they strike their roots into his corruptions and draw their nourishment from his evil life. Thus do sinners go on 'treasuring up wrath against the day of wrath and revelation of the righteous judgments of God.'

"But remember that God has not left the world in these later ages without the testimony of wrathful judgments which ought to startle and alarm the consciences of the wicked like the fires of Sodom. Let me give you what I suppose to be a true record of the fate which befell a band of bold blasphemers. In that uprising of infidelity which took place near the close of the last century there was formed at Newburg, N. Y., through the influence of a man known as 'Blind Palmer,' an association of infidels under the name of the Druidical Society. The object of the society was to uproot and destroy revealed religion. In pursuit of this object they descended to the most blasphemous mockery. At one of their meetings they burned the Bible, baptized a cat, partook of the bread and wine as appointed for the ordinance of the Lord's Supper, and gave the elements to a dog. Then the wrath of God broke out upon them. 'On the evening of that very day he who had administered the mock sacrament was attacked with a violent inflammatory disease; his inflamed eyeballs were protruded from their sockets; his tongue was swollen, and he died before morning in great bodily and mental agony. Dr. H——, another of the same party, was found dead in his bed the next morning. D—— D——, a printer who was present, three days after fell in a fit, and died immediately. In a few days three others were drowned. Within five years from the time the

Druidical Society was organized all the thirty-six original members—actors in the blasphemous ceremonies spoken of—died in some strange or unnatural manner. Two were starved to death, seven were drowned, eight were shot, five committed suicide, seven died on the gallows, one was frozen to death, and three died, the record says, *accidentally*.' Be sure of this: God has not left the world nor forgotten his judgments against his enemies, neither is he tied up and hampered by the laws of Nature. 'God is angry with the wicked every day. If he turn not, he will whet his sword: he hath bent his bow and made it ready. He hath also prepared for him the instruments of death.'

"But remember, also, that God does not limit his expression of wrath to these natural agencies. The smile of God beams direct upon the soul as the warm rays of the sun fall upon the cold earth, and the frown of God throws a shadow which darkens the soul with the gloom of eternal death."

This discourse stirred the mind of Mr. Hume in a wonderful manner. The story of God's judgments upon wicked men and dissolute cities he had read many a time in his boyhood, but the rapid review of them by Mr. Wilton seemed to bring them up with a lifelike vividness. And that view of the forces of Nature, as allied with the moral laws of God to work out wrath upon evil-doers, was new to him, but his own mind quick as thought suggested many more illustrations than Mr. Wilton had time to give. He remembered that all manner of vices—drunkenness, lust, devotion to gay, sensual pleasures—bring ruin to men. He had noticed that the saddest faces are those of worn-out lovers of pleasure, and he knew that lovers of pleasure are very quickly worn out—that five years of sensuality will waste the powers of life more than fifty years of good work. He knew also that infidels and blasphemers, whatever else they might be, were unhappy men, and died joyless, foreboding deaths. He was not exactly angry, but his heart rebelled against thus being held by the mighty power of God, willing or unwilling, and against the thought that even Nature herself had conspired against him. It seemed to him hard that he was born into such a world, and that there was no escape from it. He did not consider at the moment that God and his works were against him only because he was against

God, and that by submitting to God in loving obedience all the forces of God's world and God's providential government would turn in his favor —"that all things work together for good to them that love God."

At length better thoughts came to him. "I must know," he said to himself, "whether these things are so. I have never examined the subject to discover the truth, but have tried to find reasons for disbelieving the Bible and denying the gospel. I ought to look at the other side. If Nature and Nature's God have blessings in store for the willing and the obedient, why should not I know this and receive my share?"

Under the impulse of thoughts like these he formed the sudden resolution to join Mr. Wilton's Bible class—that is, if he would receive him willingly, of which he had no small doubt. Coming directly forward at the proper time, he said to Mr. Wilton:

"I have learned what your class is studying, and should like, I hardly know why, to join your class for a few Sundays, if you are entirely willing."

Mr. Wilton, of course, did not know the exact

state of Mr. Hume's mind; he did not know but that he came with a contentious spirit to bring up objections and propose hard questions; but he felt certain that, whatever his state of mind, the Spirit of God was bringing him to take this step. He had prayed for him; in prayer his soul had travailed in pain for him; and he felt that by way of the throne of grace he had obtained a hold upon Mr. Hume—that the Holy Spirit had bound a cord between them which could not be broken. He believed, therefore, that, whether he came penitent or angry, good would result from his coming. He gave him, therefore, a hearty welcome.

"I am not only willing," he said, "but very glad, to have you come; and as I know that you have kept yourself informed of the latest phases of modern science, I hope we shall have your help in unfolding the subject which we are engaged in studying. I think you will be able to do us good."

"Your kind welcome ought certainly to incline me to do anything which I can to help the interest of your study, but I only ask the privilege of sitting with your class as a silent listener."

The Sunday-school opened as usual, and the classes entered upon their work.

"You have come in, Mr. Hume, at just the proper point in the progress of our lessons," said Mr. Wilton. "We have been preparing the way by a brief review of the laws of heat. We have gone over the effects of heat; the conduction, radiation, and convection of heat; thermal reflection, absorption, and transmission; specific and latent heat. We have tried to form a conception of the existence and operations of heat according to the dynamic theory that heat is a mode of atomic motion. This review would have had little interest to you. We are now prepared to look at the goodness and wisdom of God in the management of heat. We are not trying to prove the existence of a Creator and Governor—we are only looking at the mighty and wise works of that God in whom we already believe. We shall find the works of God planned and wrought out with wondrous skill, and that wonderful skill is employed in the interest of goodness. God has planned and wrought for the benefit of his creatures. His wisdom and goodness are exhibited on the grandest scale and in gigantic proportions. This is all that is needed practically to demonstrate the existence of God. A good conscience does the rest. Being once assured that there is a Creator, a good conscience leaps to the conclusion that we ought to obey and serve him. Nay, the very work and existence of a conscience implies a divine Lawgiver and Ruler. To a good conscience a God is a necessity. But as we are not now attempting to show that there is a God, but to study his works, we will pass this point.

"With respect to the subject before us, let us first notice that heat is a necessity to the world and to man, and that God has made ample provision for that need. What the condition of the world would be without heat we can only conjecture. In the polar regions a natural temperature of seventy degrees below zero has been observed. At this temperature all the water upon the globe would turn to ice hard as adamant; all vegetation would cease, and with the disappearance of vegetable life all animal life must perish. The whole earth would be a frozen, lifeless, silent waste in the midst of silent space. Some lines in Byron's picture of universal darkness would fitly describe the state of the world:

> 'The waves are dead, the tides are in their grave,
> The winds are withered in the stagnant air,
> And the clouds are perished.'

This description would be no figure, for motion as well as life depends upon heat. Yet seventy degrees below zero is but the beginning of cold. 'By mixing liquid protoxide of nitrogen with bisulphate of carbon in a vacuum, M. Natterer produced a temperature of two hundred and twenty degrees below zero.' At this temperature some of the so-called permanent gases—as carbonic acid, chlorine, and ammonia—can be compressed into liquids, and it is believed that in the complete absence of all heat all the gases would become solids. But by the agency of heat the world teems with active life. Vegetation clothes the earth with a garment of beauty; and earth, air, and sea swarm with living creatures full of enjoyment. This great need of the world is bountifully supplied. The power and wisdom of God are employed in producing happiness.

"This, however, is but a part of the benefit which heat confers upon the world. The chief inhabitant of the earth is man, and man was created for something higher than bare existence. He was created for civilization and culture. The savage state is not, as some self-styled philosophers dream, the natural state of man. Nothing is so much against Nature. The natural state is that condition in which he attains the fullest development. Let a brute be placed in so unfavorable conditions that his growth is dwarfed and his natural instincts are not called into exercise, and no one would look upon that as a natural state. But man, wild, uncultured, undeveloped, is spoken of as being in his natural state. There could be no greater mistake. Culture and civilization are according to Nature, but culture and civilization require that man should get the mastery of Nature and subdue her forces. Till man gets the victory over the forces of this rough world, he spends a precarious existence in a hard struggle to gain a meagre support for his animal life. But when once science brings art, and the mastery of Nature is gained, man can rise into culture and beauty. Opportunity is given for development. He blossoms into greatness and strength. Ideal and spiritual ends take the place of mere subsistence.

"But by what agency does man achieve the mastery of Nature? By the agency of heat.

By the aid of heat man subdues the world. Heat brings the lustrous metal from its native ore; heat fashions the metal into a thousand shapes for the use of men; heat reigns as king in the curious processes of the chemist's laboratory, and the laboratory is the mother of all those modern arts which bless and beautify human life. By heat man prepares his food; by heat he drives his machinery; by heat he outstrips the flight of the winds; by heat he turns winter into summer and in his own dwelling makes for himself a perpetual springtime. For these purposes of human comfort and culture, God has provided generous stores of heat and placed them under man's control. He has placed in man's hands the means by which he can generate a heat which devours the hardest metals like stubble and a cold greater by far than Nature ever produces. We see that the Creator has provided for man as a being susceptible of culture and development, as a being of soul and sentiment, of spirit and aspiration. God has fitted the world to be the dwelling-place of spiritual beings like man."

"I beg your pardon," said Mr. Hume at this point, "that the first word I speak in your class should be a question which amounts to an objection."

"I shall be glad," said Mr. Wilton, "to hear your question, even though it be an objection. I will also answer it if I can"

"I wished to ask why it is, if God designed to provide for man's wants, that man can supply his wants, especially his higher wants—the wants of his intellectual and spiritual nature—only with the greatest difficulty and toil? The brutes supply their need with comparative ease, but man with boundless thought and labor."

"Your question is an important one, and deserves an answer. For myself, I look upon the fact to which you refer as one of the many points in which this world is adapted to human needs. Man is put in a condition which requires boundless thought and toil for the supply of his higher wants just because he possesses a nobler nature and such thought and exertion are needed for its development. Which is the more desirable condition for a young man to be placed in—one in which his every wish is anticipated and his every aspiration is gratified without exertion on his own part, or one in which opportunity and means are furnished for self-help, one in which he can supply his wants and satisfy his aspirations only by the exercise of his best abilities? Which will encourage the larger manliness and nurture the higher culture and strength? He who has no need for exertion rises at best only to a soft and feeble luxury, without mental vigor or moral force. What does man need besides scope and reward for exertion? Effort and struggle are necessities of our nature. This is especially true of man's higher faculties. Human greatness and goodness are not created by a word: they must be developed by exertion. For this reason God has made exertion necessary, and as much more necessary with man than with the brutes as his culture is more the result of voluntary, intelligent exertion. Does this explanation seem to you satisfactory, Mr. Hume?"

"I have no fault to find with it; I must think of it."

"Very well, then; if no other one has a question to ask, we will look at another subject. We will survey the storehouses of heat which God has prepared for warming the earth. Samuel, you may name the first great source of heat."

"I think, sir, that the sun is the chief source of heat."

"We certainly receive the larger part of our heat from the sun. No one can doubt this. So much of our heat comes from the sun that the temperature of the earth varies according to the sun's heat, as if that were the only supply. If but a fleecy cloud pass between the sun and the earth, we feel a decided change of temperature. A few hours less of sunshine each day, and a few degrees more of inclination to the sun's rays, change summer to winter and make the difference between the torrid and the frigid zones. Withdraw the heat of the sun altogether, and the whole world would become a desert of frozen death."

"What is the cause of the sun's heat?" asked Peter.

"You have asked a question which I cannot answer, and which no man can answer. The most careful and patient observations have been made to discover if possible the constitution of the sun; learned and curious conjectures have been brought forward to explain the source of its heat; but the positive results have not been very large. It is certain that the sun is a globe revolving upon its axis in a period of twenty-five days, nine hours, and thirty-six minutes. This is known by the motion of dark spots upon its surface. The appearance of the sun as seen through a telescope is that of a globe of fire, its surface often in a state of violent agitation and flecked here and there with dark, irregular, changeable spots. These spots are sometimes of enormous dimensions—thirty thousand or fifty thousand miles in diameter. They present a dark centre with a narrow border or penumbra of lighter shade. To account for these spots, it has been conjectured that the body of the sun is dark, but surrounded by a double envelope of clouds, the outer layer of which is intensely luminous. Openings in such enveloping clouds would present an appearance like the spots upon the sun. According to this supposition, the heat and light of the sun proceed, not from the body of the sun, but from this luminous enveloping cloud. But granting that this supposition is true, it gives no explanation of the origin of the sun's heat. Laplace conjectured that the sun is a globe of fire in a state of violent, explosive conflagration, and that the spots are enormous crater-like caverns in its surface. Newton conjectured that comets falling into the sun and being consumed feed the solar fires and maintain its temperature. The reception of the dynamic theory of heat has led to the revival, in a modified form, of this conjecture of Newton. It is suggested that meteors or meteoric matter falling into the sun generates its heat by the force of concussion. To show that the intense heat of the sun might be thus generated, elaborate calculations have been made. It has been demonstrated that if the sun were a solid mass of anthracite coal, its combustion would maintain its heat at its present rate of emission only five thousand years, while the falling of the planet Jupiter into the sun would generate an equal amount of heat for thirty-five thousand years. A lump of coal falling from the earth to the sun would produce three thousand times more heat by the concussion than by its combustion.

"The nearest approach that has been made, of an exact and scientific kind, toward determining the constitution of the sun's surface has resulted from an examination of the *solar spectrum*. A ray of light, by passing through a triangular prism of glass, is, as you know, divided into its elements, or constituent colors. The ray of light is spread out like a half-open fan. This divided and expanded ray, thrown upon a screen, is called the spectrum. An examination of the solar spectrum by a microscope shows certain fine dark lines across it. The lines are invariably the same in their position and grouping. The spectrum of the stellar light is found to differ from that of the solar light, and the light of one star differs from that of another star. Light from incandescent metallic vapors gives bright lines across the spectrum. Each metal has its own number, position, grouping, and color of these spectral lines. By comparing the solar spectrum with the spectra of the various metals—the processes are curious and the explanation difficult to be understood—corresponding lines are discovered, and the conclusion is reached that the sun's atmosphere contains the vapors of several of our well-known metals, as iron, nickel, sodium, potassium, and others. This is a most curious and marvelous scientific feat, to make an approximate chemical analysis of the sun and stars by means of their light. The conclusions, however, seem trustworthy.

"Can you tell us, Ansel, whether the earth receives heat from the moon and stars?"

"I cannot, sir."

"I should be glad, Mr. Hume, to have you instruct us upon this point."

"In regard to the fixed stars," answered Mr. Hume, "counting them as the remote suns of other planetary systems, we must believe that they radiate more or less heat upon the earth; some indeed have extravagantly maintained that we receive from them nearly as much heat as from the sun. The heat received from them is so small that we perceive no difference whether they be hidden, or shine with their utmost brilliancy. I do not know that investigations have been made to determine scientifically their exact thermal influence upon the earth. But little more can be said about the heat of the moon. The light of the full moon, concentrated by a two-foot burning-glass and thrown upon the bulb of the most delicate thermometer, produces no perceptible effect. By means of the electroscope or galvanometer, it is said, however, that the moon's heat has been detected. At a late scientific convention held in Chicago, Prof. Elias Loomis read a paper, in which he stated that Mr. Harrison of England, by a comparison of observations made for sixteen years at Greenwich, nine years at Oxford, and sixteen years at Berlin, has discovered that the moon exerts a sensible influence upon the temperature of the earth, the highest temperature occurring from six to nine days after the new moon and the lowest about four days after the full moon. The conclusion, the opposite of what we should naturally expect—the higher temperature occurring when the enlightened face of the moon is turned from the earth—was explained by supposing the moon's heat to be dark heat which would be absorbed by the vapors and the clouds, and thus tend to warm and dissipate them. By the dispersion of the clouds, the radiation of heat from the earth's surface would go on more rapidly and the temperature would fall. According to this explanation, the lunar heat reduces instead of raising the temperature of the earth. The difference of temperature due to the moon's influence Mr. Harrison believed to be two and a half degrees. Upon extending his calculations through forty-three years of observations made at Greenwich, he found the difference reduced to about one degree. As for myself, I confess myself still a skeptic touching the supposed influence of the moon upon temperature."

"Upon that subject, I think," said Mr. Wilton, "that we must wait patiently for more light. The popular superstitions which refer sickness and health, and every kind of good or evil fortune, to the benign or malignant influence of the moon, we, of course, must reject. Samuel, will you name the second chief source of heat?"

"I am obliged to answer as Ansel answered just now—I cannot tell. The enormous amount of wood and coal burned amounts to something, but this can have very little effect upon the temperature of the earth."

"The second great store of heat is the internal heat of the earth," said Mr. Wilton. "The importance of this store of heat we can easily understand by considering that the earth is a mass of molten mineral matter cooled and hardened upon the surface. The crust upon which we live is warmed from beneath by an ocean, or rather a globe, a world, of glowing molten rock. Deep excavations have been made in mining operations, and artesian wells have been bored to still greater depths—as deep as two thousand, three thousand, or thirty-five hundred feet. The heat of the sun penetrates not more than seventy-five or a hundred feet; below that depth the temperature of the earth remains the same throughout the year. Below the point of constant temperature the heat of the earth is found to increase regularly and constantly. The rate of increase varies in different regions, but the average rate is about one degree of temperature for each fifty or sixty feet of descent. From this rate of increase it is easy to calculate the temperature at any given depth. At a depth of less than two miles water would boil. At twelve miles in depth the rock becomes incandescent. At twenty-two miles silver melts, at twenty-four miles gold melts, and at thirty-five miles cast iron becomes liquid. Volcanic eruptions also demonstrate the existence of immense masses of molten rock in the interior of the earth; and we can account for the existence of volcanoes only by supposing that they now communicate or once communicated with the deep interior heat of the earth. The thickness of the earth's crust is, however, a matter of conjecture. The melting point of different substances rises as the pressure upon them increases, and as the density of the rock increases its conducting power becomes greater. The crust of the earth, therefore, may be fifty miles in thickness, or it may be one hundred miles or two hundred or three hundred miles. The effect of this internal heat in maintaining the temperature of the earth must be very great."

"I want to ask," said Peter, "how this internal heat came to exist, and how it is maintained?"

"This, like your former question, is altogether beyond our knowledge. All that we certainly know is that God made it thus. The process of creation, if indeed God did not create the earth by a word, without a process, is a matter of sublimest and most venturesome conjecture. According to the opinion of some, the elements of which the earth is composed were created separate and uncombined, and were suffered afterward to unite by their chemical affinities. This chemical combination would be nothing else than a tremendous conflagration, and the result would be the most intense heat of which we can form a conception. Others have dreamed of a 'fire-mist' created of God and by some means condensed into worlds. The temperature of the earth is maintained, so far as we know, only by the poor conducting quality of the enveloping crust preventing its cooling. At the present rate of radiation, millions of years would be required to render the change of temperature perceptible.

"What is the third great natural source of heat? I will ask Mr. Hume."

"Mechanical action, or force transmuted to heat."

"Will you please explain this?"

"Strictly speaking," said Mr. Hume, "this is not to be counted an original source of heat. But heat is used in the production of winds and waves, the flow of rivers, and all the ceaseless activities of the world, and this force reappears from time to time transmuted again to heat. Whenever in the friction of air and of water, in the dashing of matter against matter and force against force, motion and force seem to be lost, heat is produced. The water of the sea after long storms is said to be sensibly warmed. We can appreciate the amount of heat generated in this manner only by considering in how many thousand ways force is meeting force and motion is destroyed. All this lost motion —lost as sensible motion—reappears as atomic motion, that is, as heat. Such heat has been applied to artificial uses. Heat generated by the friction of iron plates ground together has been used for heating buildings."

"And this transmutation of living force and heat," added Mr. Wilton, "is but one of many illustrations of God's economy in the management of heat. Nothing is wasted. The voices of Nature all echo the words of Jesus: 'Gather up the fragments, that nothing be lost.'

"The fourth source of heat is chemical action. What is the chief form of this which is used for the production of heat? Samuel may tell us."

"Combustion, I think, sir."

"That is right; and the most common form of combustion is the combination of carbon with oxygen. This is commonly employed, not because it generates the most intense heat, but because carbon exists so abundantly, and is the most available and the cheapest. The most common form of carbon is wood and coal. This is that storehouse of heat which God has placed in man's keeping. Without this the larger part of the earth's surface would be uninhabitable. This renders culture and civilization possible. Without it the arts could have no existence. The key of this storehouse of heat God has given to man, so that he may enter in and use its treasures at his pleasure. In the finer arts where very great heat is required, hydrogen is used in place of carbon. Jets of oxygen and hydrogen gas thrown together constitute what is called the oxy-hydrogen blowpipe, and generate the intensest heat which can be produced by man.

"Another source of heat not often mentioned is electrical force. This, like mechanical force, may be transmuted into heat. An electric current sent through an insufficient or poor conductor heats it, and, if the current be sufficiently strong, consumes it. Thus lightning-rods are sometimes melted and buildings set on fire.

"These, then, are the natural reservoirs of heat: 1, the sun and other heavenly bodies; 2, the internal heat of the earth; 3, living force, or motion; 4, chemical action; 5, electric force.

"We can hardly over-estimate the abundance of these natural supplies of heat. The world is warmed on the most munificent scale. The earth receives from the sun heat sufficient to boil three hundred cubic miles of ice water per hour, and the whole sum of the sun's heat would boil 700,000,000,000 cubic miles of ice water in the same time, that is, the heat radiated by the sun would boil a mass of ice water of the size of our globe in twenty-five minutes.

"The amount of carbon provided by the Creator is enormous beyond conception. Vast regions of country are covered with dense forests, but the fuel from the forests is but a handful in comparison with the fuel stored up in coal-beds below the surface of the earth. Mr. Mitchel estimated the extent of the coal-beds of a portion of Europe as follows: Great Britain, 12,000 square miles; Spain, 3500; France, 1700; Belgium, 5180. Mr. R. C. Taylor has made a like estimate for North America, giving to British America 18,000 and to the United States 134,000 square miles.

"These estimates, you will notice, say nothing of Asia, Africa, South America, or the islands of the sea, and include only the smaller part of Europe. In the United States, also, new coal-fields are constantly discovered. The supply of carbon for fuel seems exhaustless. In the British islands about 100,000,000 tons of coal are mined annually. At this rate the known supply would last for a thousand years. In the United States the supply has no known limit.

"You will keep in mind that this supply of heat is also a supply of mechanical force. The coal-fields are an exhaustless storehouse of heat and power. They warm the dwellings of man and drive millions of engines working with the strength of Titans for human welfare.

"In this bountiful supply of heat to warm the earth and serve human needs must we not see a kind design on the part of the Creator? God has provided that which the world needs. He has provided without stint or limit. The general heating of the globe he accomplishes by his own power. He has provided for human culture, development, and happiness by placing stores of heat under man's control. He has furnished scope and means and encouragement for achieving greatness and goodness. He has put man in the condition which a wise father would desire for his son.

"In our next lesson we will look at the preservation and distribution of heat, some of the primary elements and arrangements upon which the temperature of the earth depends."

CHAPTER VII.

PRESERVATION AND DISTRIBUTION OF HEAT.

ANOTHER Lord's Day comes, and the members of the class are, as usual, all in their places. They find the subject increasing in interest after leaving the review of the laws and principles of heat.

"A week ago," said Mr. Wilton, "we looked at the chief sources of heat. These are the sun, the internal heat of the earth, chemical action, in which combustion is most important, electrical action, and mechanical action, or 'living force.' The amount of heat furnished from these sources is above all comprehension. The Creator seems bountiful even to prodigality in supplying heat for the needs of the world and the uses of man. But with all this largeness of supply the provision would prove wholly inadequate if it were not prudently husbanded

CURIOSITIES OF HEAT. 153

and all the avenues of waste carefully closed. Men of ample incomes sometimes come to want from too free expenditure. Their incomes are large, but their expenses are larger. So it would prove in respect to heat if Nature were not as prudent in saving as she is bountiful in providing. Will some one mention some of the general methods by which the waste of heat is prevented?"

No one answered. Mr. Hume did not think it best to put himself forward in answering questions, and therefore answered only when personally addressed. The others were silent because they had nothing to say.

"I see that I shall have to suggest the answer. Ansel, what part of the atmosphere is warmest?"

"The bottom, I suppose, for the higher a man goes up upon the lofty mountains or in a balloon, the colder he finds the air."

"That is right; and we need to ascend only about three miles, even in the tropics, to reach the region of perpetual snow, while in the polar regions the line of perpetual freezing comes down to the sea level. What would be the effect, Ansel, if the atmosphere were as warm, or warmer, at the top than at the surface of the earth? How would that affect the rate of radiation from the earth?"

"It must, of course, increase the radiation very much. With the temperature twenty or fifty or seventy degrees below zero, the radiation must be very little."

"By some means, then, the atmosphere is kept warm at the surface of the earth and cold in the higher regions, and in this manner the radiation of heat into open space is prevented. This is accomplished notwithstanding that the top of the atmosphere is nearer the chief source of heat, the sun. This would be no very easy problem if its solution were left to human ingenuity. The explanation is very simple, however, when once suggested. The atmosphere is diathermic, that is, it permits the luminous heat from the sun to pass directly through it without heating the air, but the solid earth stops the heat by absorption, and is warmed by it. The warm surface of the earth imparts, in turn, its heat to the atmosphere resting upon it. This warm air, being expanded by the heat received, becomes lighter than the cold air around, and rises, or rather is forced, upward by the greater weight of the colder air. But as it rises and the pressure of the air is diminished it expands still further. By this expansion its sensible heat becomes latent, that is, the heat is transmuted into force, and, as force, is incapable of being radiated. In this manner radiation from the upper surface of the atmosphere is greatly hindered and waste of heat is in a good degree prevented.

"In respect to this heating of the atmosphere from the surface of the earth, a layer of clouds sometimes forms a kind of second surface which receives the sun's rays and warms the air above. A few years ago I saw a balloon ascension in Providence, R. I. The day was bleak and chilly, and the sky entirely covered with clouds. The aeronauts were expecting a chilly voyage. The balloon shot like an arrow toward the zenith, and in five minutes was completely hidden by the clouds. But to the surprise of the voyagers of the sky, on passing through the clouds their thermometer rose ten degrees. This, doubtless, must very often be the case. The air above the clouds must often be warmer than that below.

"I think you all must have noticed illustrations of this principle on a small scale. Have you not seen that snow and ice often melt around straws and sticks, the snow or ice remaining still frozen at a little distance, as if the sticks and straws were warm and had melted them? Have you not seen a dark-colored board covered with ice, and the ice remain firm till the sun shone upon it, and then the ice melt upon the under surface, leaving the upper surface unaffected?"

"I have seen such things a great many times," said Peter, "and wondered what the reason was."

"The reason is that ice is *diathermic*. Heat passes through the ice without warming; but when the rays of heat fall upon the stick or stone or board, the heat is absorbed, the dark body is heated and in turn warms and melts the ice. In the same manner the atmosphere is warmed. The heat-rays of the sun pass through the atmosphere and fall upon the surface of the earth; the earth is warmed, and in turn warms the air resting upon it.

"The gases and watery vapor contained in the air also hinder the radiation of heat from the earth. Pure atmospheric air is perfectly diathermic to both luminous and dark heat, and vapors and gases are also diathermic to luminous heat. But to dark heat some of the gases are almost impenetrable. Ammonia stops dark heat almost completely. In a smaller degree watery vapor does the same. Gases and vapors thus serve as blankets to keep the earth warm. The heat of the sun, being luminous heat, penetrates the atmosphere with its vapors and foreign gases, and falls upon the earth almost without loss, but, being absorbed by the earth, it becomes dark heat, and cannot be radiated back through the same gases and vapors. Vapor serves thus as a valve: it admits the heat of the sun to the surface of the earth, but prevents its escape. Prof. Youmans calls watery vapor the barb of heat; it catches the heat of the sun and holds it fast.

"Who can sufficiently admire the simplicity of these arrangements for preventing the radiation of heat into the stellar regions?—and their efficiency is no less admirable than their simplicity. Arrangements like these show that the Creator had a definite object in view, and that object is benevolent. For the advantage and enjoyment of the inhabitants of this world these arrangements were made.

"We ought at this point to look at those adjustments by which the earth receives just the amount of heat needed to maintain the requisite temperature. The importance of maintaining some certain average temperature cannot be over-estimated. Every animal and plant has its own *habitat*—that is, its natural dwelling-place or location—outside of which it perishes or maintains a stunted and precarious life. The habitat of animals and plants depends in a very great degree upon temperature. What a panorama would be seen if we could fly like a bird from the equator to the poles, and look down upon the ever-changing animal and vegetable life as we pass! How the luxuriant vegetation and flaunting colors of the tropics would shade off into the scantier vegetable life and more sober hues of the temperate zones, and these in turn die out and disappear in polar barrenness! We should see the lion and tiger give place to the bear and the wolf, the elephant and camel to the ox and horse, and these to the white bear and reindeer. This sublime panorama we see, in miniature, in ascending lofty mountains in the tropics. Around the base of the mountain flourish the rich and various productions of the torrid zone; a few thousand feet of elevation bring us among the productions of the temperate zones. The most valuable fruits and grains thrive. Then vegetation becomes scanty and stunted, and at last disappears. The top of Mt. Washington, 6234 feet high, in latitude forty-four degrees, is as bare of trees and plants and every form of vegetation as the north pole.

"The fitting temperature is almost as necessary to the animal tribes as to vegetable life. Animals which are native to the tropics do not thrive in colder countries, or if the difference of temperature be very great, they perish. A change from a cold to a warm region is equally disastrous. Man indeed transfers animals from their natural habitat by protecting them from the extremes of temperature, but this is, of course, no exception to the general principle of which I am speaking. A change of only a few degrees in the mean annual temperature would render this earth a hard place for even the human race to subsist. But the temperature of the earth depends upon many a wise adjustment—how many, we cannot tell. Will you tell us, Samuel, the first adjustment or arrangement upon which the temperature of the earth depends?"

"It must depend chiefly I think upon the intensity of the sun's heat."

"Whether or not that be the chief adjustment by which the right temperature is secured, it is at least a very important item. The intensity of the sun's heat must, of course, be considered in connection with its distance from the earth. The distance of the sun is no less important than the power of his rays; indeed, in one sense, it is more important, for if the intensity of the sun's heat were doubled, the temperature of the earth would be increased only twofold; whereas, if the earth were brought to one-half its present distance from the sun, the heat would be increased four times. Heat being one of the radiant forces, its intensity diminishes in proportion to the square of the distance through which it acts. If the earth were 190,000,000 of miles from the sun instead of 95,000,000, as it now is, the force of the sun's rays would be diminished fourfold. The Creator has so fixed the distance of the earth and sun, and the power of the sun's heat, as to give to this world a temperature suited to its various inhabitants.

"The temperature of the earth has also some dependence upon our atmosphere. Can you tell us, Ansel, how the temperature of the earth is affected by the atmosphere?"

"You have already told us that the atmosphere is *diathermic*, allowing the heat of the sun to fall upon the earth almost undiminished in force. If the air were so constituted as to intercept the sun's rays, it is plain that the earth would receive less heat."

"This adaptation of our atmosphere to transmit the sun's rays," said Mr. Wilton, "is more subtle than it appears at first sight. It is not merely a matter of depth and density, though those are important considerations, nor is it merely a question of the elements of which the atmosphere is composed. Simple gases are *diathermic*. The atmosphere is therefore made up of two simple gases, oxygen and nitrogen, not chemically combined, but mixed together. Compound gases intercept the passage of heat. Ammonia, composed of hydrogen and nitrogen chemically united, almost wholly stops it. Even ozone, which is nothing but oxygen in a changed or *allotropic* state, is not *diathermic*. The *diathermic* quality of the air depends, then, not only upon the fact that it is composed of simple elements mingled, but not chemically joined, but also upon the *state*, or *condition*, of those simple elements.

"Another point deserves attention. Oxygen is an element having a wide range of very strong and active affinities. It is ready to unite with every known substance, fluorine excepted. What if some other equally active element were mingled with oxygen to form the atmosphere? What if, in place of nitrogen, vapor of sulphur were substituted? What if hydrogen were put in the place of nitrogen? The two elements would combine in sudden combustion or explosion, and the atmosphere itself would perish. But nitrogen is a substance so sluggish and inert that it can be brought into union with oxygen only by indirect processes. Because the air is composed of one so inert element as nitrogen, the atmosphere is preserved, and, what is almost as important, it is kept, as it now is, composed of simple elements, and hence *diathermic*. If our atmosphere were a compound gas, the world would perish with cold.

"The temperature of the earth depends also upon certain qualities of the earth's surface. I should be glad to have Mr. Hume explain this."

"I suppose," answered Mr. Hume, "that you

refer to the qualities of the earth as an absorbent and conductor of heat. The earth must needs have the capacity of receiving and retaining the heat which falls upon it from the sun. If the earth's surface were polished and brilliant, the heat of the sun would be reflected into space as from the surface of a mirror, and very small advantage would the earth receive from the solar heat. A dark soil absorbs heat more readily than a soil of lighter color, and a wet soil, on account of the high specific heat of water, requires more heat to raise its temperature than a dry soil. The mineral elements of the soil and its compactness or porosity also help to make up its capacity for receiving and retaining heat. The color and constitution of the soil sometimes go far toward making the climate of a region. The conducting qualities of the earth's crust in its profoundest depths also must be taken into account. If the crust of the earth were composed of silver, or any other substance of like conducting quality, and the interior of the earth were molten rock, as it now is, the interior heat would be so rapidly conducted to the surface that everything upon the earth would be consumed."

"Upon so many circumstances wisely adjusted

and nicely blended," said Mr. Wilton, "does the temperature of the earth depend. The intensity of the sun's heat, the dimensions of the earth's orbit, the constitution of our atmosphere in the subtlest qualities and relations of its elements, and the material, structure, and color of the earth's crust,—on all these and many other things which I cannot stop to mention depends the temperature needful for the well-being of the inhabitants of this globe. I beg your pardon, Mr. Hume, but allow me to ask whether such a combination of agencies and conditions, uniting to work out good for man, does not seem to you quite superhuman and worthy of a wise and good Creator?"

"I cannot deny it, sir," he replied; "I am not prepared to make any objections. There are many things painful to man in the vicissitudes of heat and cold, and if I were to make a world, I suppose I should leave them out, or perhaps make the world upon a very different plan. But I am not prepared to affirm that any changes which I could make would be improvements, though I have thought until recently that more of knowledge and power, and perhaps more of chance, too, than of wisdom and goodness, were displayed in the works of Nature. But I must confess my opinion has been much modified."

"I think your change of mind is in the right direction, and I am glad that it is so. We learn the secrets of Nature and appreciate her spirit much better when we come as reverent questioners than when we come with preconceived notions and a patronizing air. I can well understand your feelings and state, for I myself have traveled over the same ground. My eyes were once dazzled with the glories of science; I worshiped at the shrine of natural laws. But I have learned that God is greater than Nature, the Creator is mightier than the creation. Nature has no mind or purpose apart from the plan and will of the supreme Architect and Ruler, and this inner plan and purpose of Nature is seen only in the government and discipline of our sinful race. I shall greatly rejoice for you and with you if you shall go on to the same end which I have reached."

"I shall much rejoice if I reach some satisfactory and peaceful conclusion."

"To understand the management of heat," said Mr. Wilton, "we must take note of the differences and fluctuations of terrestrial temperature. The sources of heat are constant. The sun sends out its flood of heat uninterrupted and changeless for ever. The internal fires of the earth give an even inward heat. Mechanical and chemical agencies are active everywhere. These sources of heat do not fluctuate, flaming up and dying away, yet temperature is the most variable of all inconstant things. In passing from equator to pole we go from torrid to frigid, from everlasting summer to everlasting winter. And not only this, but in the same region the temperature never remains the same for even twenty-four hours. The thermometer may pass from forty degrees above to thirty below zero in a very few hours. We must first consider the agencies by which these inequalities are produced. Ansel may mention the first of these."

"The shape of the earth," said Ansel.

"How does the form of the earth operate to produce inequality of temperature?"

"The earth is a sphere, and the rays of the sun fall upon it in nearly parallel lines. Upon the centre of the hemisphere which is turned toward the sun the rays fall perpendicularly, the sun is directly over head, while toward the edges

CURIOSITIES OF HEAT. 167

of the hemisphere, on account of the curvature of the earth's surface, the rays fall more and more slanting, as if the sun were sinking toward the horizon."

"What is that inequality of temperature which is produced by the shape of the earth?"

"The five zones," answered Peter.

"This subject is so well understood," said Mr. Wilton, "that I need not spend time in explaining it. Every boy knows the difference between setting his wet slate before the fire to dry so that the heat will fall squarely and perpendicularly upon it and placing it edgewise to the fire. Upon the torrid zone the sun shines perpendicularly, upon the temperate zones obliquely, and upon the frigid zones still more obliquely, and during a part of the year the sun is entirely hidden. In proportion as the rays of heat fall obliquely, any given amount of heat is spread, so to speak, over a larger surface, and the larger the space over which it is spread, the feebler it becomes. What is another cause of inequality of temperature?" No one answered. "Samuel, what is the cause of day and night?"

"The turning of the earth upon its axis."

"And the rotation of the earth upon its

axis," continued Mr. Wilton, "brings not only an alternation of light and darkness, but also of heat and cold. The heat of the sun is withdrawn along with the light. The heat of the sun is not withdrawn from the earth, but one-half of the earth's surface is constantly turned away from its influence. This must produce a daily change of temperature. This diurnal fluctuation of temperature may be very small or it may amount to seventy or eighty degrees. Samuel, what is a third cause of unequal temperature?"

"The inclined position of the earth's axis and the revolution of the earth around the sun cause the change of seasons."

"If it were not for this, the earth would still have her zones of seasons; a part of the earth would have endless summer, a part endless spring, and the rest unbroken winter, but the alternation of seasons at the same place would be unknown. The axis of the earth is now inclined about twenty-three degrees, twenty-seven minutes, twenty-three seconds to the plane of the earth's orbit, and as this axis maintains constantly the same position, being parallel in one part of the earth's orbit to its position in any other part of its orbit, during one part of the year the north pole is turned twenty-three and a half degrees toward the sun, while in the opposite part of the year the south pole is in like manner brought into the light and heat. This causes the sun to appear to move to and fro, north and south, twenty-three degrees, twenty-seven minutes, and twenty-three seconds from the equator in either direction. The tropics, or turning-places, mark the limits of the sun's northern and southern journey. Everywhere between the tropics the sun, at some period of the year, passes through the zenith, that is, exactly overhead at noon. North and south of the tropics the sun seems to rise higher in summer and to sink lower in winter. In summer the sun at midday is about forty-seven degrees nearer the zenith than in winter. Within the polar circles, which are the same distance from the poles as the tropics from the equator, the heat of the sun is entirely withdrawn during a portion of the year, and during another portion of about equal length the sun does not set. The extremes of temperature, caused by the inclination of the earth's axis and its revolution around the sun, are very great. In the northern part of Minnesota, the temperature rises in summer to one hundred degrees, and in winter sinks to fifty degrees below zero, giving thus an alternation of one hundred and fifty degrees.

"In this connection you may also remember that the sun is nearer the earth in one part of its orbit than in another part. This difference amounts to about 3,000,000 miles. The sun also remains eight days longer north of the equator than south of it. Our summer, therefore, is eight days longer than the summer of the southern hemispheres, and our winters are correspondingly shorter. These differences tend, however, to balance each other, for while the southern summer is shorter, the sun at that time is nearer, and while our summer is longer, the sun is more distant. Peter, you may explain to us the effect upon temperature caused by the division of the earth's surface into land and water."

"I learned while studying physical geography that the temperature is more even upon the sea than upon the land. But why, I do not know."

"The smooth surface of the sea reflects heat better than the rough land: for this reason, a larger proportion of the heat which falls upon the sea is not absorbed, but reflected and lost, so far as the temperature of this world is concerned. Water is also a very poor conductor of heat, and has withal a very high specific heat. For these reasons the sea receives and parts with heat more slowly than the land, and its absorption or radiation causes a smaller variation of temperature. The result is, therefore, that the sea is cooler in summer and warmer in winter than the land, and the average ocean temperature is lower than the mean continental temperature. The land receives heat more readily and parts with it more rapidly; the fluctuations of temperature must therefore be greater. Hence, the interiors of the continents have much greater extremes of temperature than the sea-board. But of the influence of water in equalizing temperature I shall have occasion to speak again more at length, and will pass it by for the present. What effect, Peter, has the unevenness of the earth's surface upon temperature?"

"The higher we ascend upon mountains, the colder we find it."

"That is, Peter, the greater the elevation of

any place or country above the sea level, the lower the temperature. Almost the whole surface of the earth is an alternation of mountain and hill, valley and plain. One continent has a very much greater mean elevation than another. One region or tract of country lies sloping toward the sun, another is inclined from it. The effect in the one case is the same as if the sun were brought more nearly overhead; in the other case, the sun is depressed toward the horizon. It is all the same as if the region of country were brought nearer the equator or removed farther from it. The effects of the curvature of the earth are obviated or exaggerated. Do clouds tend to produce inequalities of temperature?"

"I think they must do this," answered Samuel. "Clouds cover one portion of the earth's surface and shut out the heat of the sun, while other portions are well exposed to the sun's rays."

"That is right, Samuel. Does any one think of another cause of inequality of temperature?"

There was a pause. Then Mr. Hume answered: "Considering the unmeasured cycles of the past, the gradual cooling of the earth has brought a great change of temperature."

"And this change," continued Mr. Wilton, "has been very important for the welfare of the human race. At the present temperature of the earth, the coal-beds, so necessary for the culture and progress of the race, could hardly have been formed, and at the temperature of the carboniferous periods, when the coal-beds were deposited, the human race could with difficulty have survived. The high temperature required to prepare the earth for man is now no longer needed, but would prove destructive. And this great change of temperature was doubtless caused by the cooling of the earth.

"The result of all these agencies—the shape of the earth, its daily and yearly motions, the inclination of its axis, the eccentricity of its orbit, the division of its surface into land and water, the varying elevation of its surface, and the clouds and storms that hide the sun—is that we have great extremes and rapid transitions of heat and cold, and every variety of climate. These changes of temperature are often painful and, unless guarded against, dangerous. Yet, taken as a whole, can one doubt that variety of climate and change of temperature are of advantage to man? What weariness and lassitude a changeless temperature would bring! How the cooler air of the night comes as a tonic after the relaxation of the heated noonday! Who can estimate the value of our northern winters, not alone in building up a vigorous and nervous physical frame, but in helping the culture of men and nurturing the domestic virtues? We might almost say that her winter evenings have been the making of New England. But periods of heat are needed for bringing fruit and grain to ripeness. What variety and richness of productions for the use of man the different zones furnish! The supply of man's wants would be comparatively meagre if we had but one zone, even though we had our choice of the zones. But every zone is necessary for the perfection of the temperate zones. That we may have the warmth of summer in the temperate zones we must have the torrid zone. That we may have the tonic cold of the temperate zones we must needs have the severity of polar winters. I do not mean that the Creator could not devise a world that should not have these painful extremes, yet enjoy the advantages of the temperate regions. But that would plainly require a world constituted upon principles very unlike those which now prevail.

With God this is doubtless possible, but the mode is to us inconceivable. But we can easily see that by the present arrangement of things God has secured many great advantages for man—how many and how great, we can hardly understand—and the apparent disadvantages we cannot positively affirm to be real evils. We can safely declare that this world is well adapted to man's necessities. But these inequalities of temperature are modified and softened by a most comprehensive and beneficent system of agencies by which the extremes are prevented from becoming destructive. In this system of compensating agencies two great divine ideas are clearly developed, economy in the expenditure of heat and benevolence toward man. Upon this subject we are now prepared to enter."

CHAPTER VIII.

MODIFICATION OF TEMPERATURE.

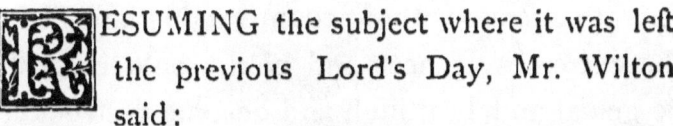ESUMING the subject where it was left the previous Lord's Day, Mr. Wilton said:

"We saw at our last session that the most prominent and permanent features of the earth tend to produce differences and great extremes of temperature. These variations of temperature within due limits must be regarded as beneficial, if not absolutely essential, to the well-being of the human race. The different zones give the world a richer and more varied supply of food, and finer and more varied plants and animals. The change of seasons gives variety in the experience of life; the warmth of summer ripens the fruit and grain, and the cold of winter tones up the physical strength; nay, the winter's frost is a natural subsoiler, loosening up the hard earth and promoting vegetable growth. As for man's higher interests, no one can tell how much the world is indebted to winter evenings, to a period of darkness longer than is needed for sleep, and a period of cold during which the work of husbandry may largely cease. Learning, the domestic virtues, and religion are greatly indebted to our winters. But were these agencies which tend to produce inequality of temperature suffered to operate without counteracting influences, the extremes of heat and cold would cease to be genial and healthful, and become destructive. We are now to begin the consideration of those counteracting agencies by which the extremes of temperature are moderated.

"Let us look first at the daily fluctuation of temperature caused by the revolution of the earth upon its axis. The rotation of the earth brings every place by turns under the influence of the sun's rays, and in turn withdraws it from the heat of the sun, thus producing a daily change of temperature. How is this diurnal change of temperature alleviated?"

This was addressed to all, but no one answered. "Mr. Hume, I should be glad to have you suggest the answer."

"There are two chief agencies," Mr. Hume

replied—"first, the absorption of heat during the day and the radiation of that heat during the night; and, secondly, the formation of watery vapor during the day and the deposition of dew by night."

"The first of these agencies," said Mr. Wilton, "is so plain that very little explanation need be made. During the day, while the sun is shining and the temperature is rising, the surface of the earth, the rocks, the trees, and all things are absorbing heat. This heat is, so to speak, laid up in store, ready for use in time of need. In due time the sun sinks low and sets behind the horizon; the supply of heat is cut off and the temperature begins to fall. Then all those objects which during the day were laying up heat in store begin to radiate heat into the air, and by their contact with it keep up its warmth. Commonly, the temperature falls so low that bodies radiate more heat than they absorb before the setting of the sun. In this process water plays a very conspicuous part. You will call to mind what was said before about the large specific heat of water. By means of this, water is able to store up heat in large amounts—larger in proportion to its weight than any other substance except hydrogen gas. The heat that is stored up during the day is given off by contact with the air and by radiation during the night.

"But water plays a still more important part in moderating the daily fluctuations of temperature by the process of evaporation and the formation of dew. Call to mind what was said of the formation of vapor when we were speaking of latent heat. Heat water to two hundred and twelve degrees—the boiling point: it must still be heated a long time before it evaporates. Boiling water must receive five and a half times more heat to give it the form of vapor than to raise it from the freezing to the boiling point; that is, about one thousand degrees of heat are required to turn boiling hot water to vapor. The same amount of heat is required for the formation of vapor whatever the temperature of the water from which the vapor rises. There is only this difference—vapor from cold water is cold, while vapor from hot water is hot. Evaporation goes on more rapidly in proportion as the temperature rises, but vapor is formed at all temperatures. Evaporation goes on from ice. The Alpine glaciers, or rivers of ice, sink away several feet by evaporation from their surface during their slow course of many years down the mountain ravines. This process of evaporation goes on, I say, during the day, and in the formation of vapor an amount of heat which would raise an equal weight of water through one thousand degrees of temperature is used up.

"This vapor which is formed is not supported *by the* air, as men commonly suppose. It is true that clouds are held up by the atmosphere, but clouds are condensed vapor—minute globules of water floating in the air. Vapor is invisible. You must have noticed that steam is invisible till it is condensed by contact with the colder air. Vapor rests upon the earth and supports itself by its own elastic force, just as the atmosphere supports itself. The presence of air makes no difference with the formation of vapor, except that in a vacuum vapor forms very much more rapidly, because no air stands in its way. But at any given temperature, in the air or in a vacuum, the same amount of vapor rises in due time, and the same amount can support itself. Vapor seems to circulate between the atoms of air, as sand fills the spaces between marbles.

At the temperature of four degrees below zero vapor equal to two-thirds of an inch of water can be formed and support itself by its elasticity; that is, the elastic force of vapor at four degrees below zero is equal to two-fifths of an ounce per square inch; at thirty-six degrees vapor equal to two and two-thirds inches of water can support itself; at eighty degrees vapor equal to thirteen inches of water can exist; at one hundred and seventy-nine degrees, seventeen feet; and at two hundred and twelve degrees nearly thirty-four feet; that is, vapor at two hundred and twelve degrees has an elastic force of fifteen pounds to the square inch. Let us suppose that at sunrise the air has a temperature of thirty-six degrees, and that as much vapor is already formed as can sustain itself at that temperature. As the sun sheds down his rays the temperature rises and more vapor is formed. We will suppose that half an inch of water is evaporated. Some of this vapor will be carried by ascending currents of air into the higher regions and condensed into clouds, some will be carried by winds into drier and warmer regions, yet the amount of vapor will increase during the day. We will suppose that during the night the temperature falls again to thirty-six degrees; all the excess of vapor above two inches and two-thirds of water will be condensed and become dew or fog, and in this condensation the thousand degrees of heat absorbed in the formation of the vapor will be given out again. If vapor equal to one inch of water be condensed, heat is set free sufficient to boil a sheet of ice water, five and a half inches in thickness, extending over the whole region; that is, it would be all the same as if a fire were kindled on every square rod of land hot enough to boil during the night more than twenty barrels of ice water. In this illustration I have supposed a larger condensation than commonly takes place, but very much less than is conceivable. Suppose that the temperature is eighty degrees, and that, as is possible, more than one foot of water exists in the state of vapor. Let the temperature fall to thirty-six degrees, and full ten inches of water must be condensed, setting free heat which would boil four and a half feet of ice water. So large a condensation as this never takes place in twelve hours, partly because the full amount of vapor which might be formed is never actually produced, and partly because the condensation of but a small part of this vapor would check the fall of temperature and prevent farther condensation. The supposition that I have made shows the possibilities of this method of moderating extremes of heat and cold. Were it not for these processes, our days would be much warmer and our nights much cooler than they now are. By the formation of vapor the excess of heat during the day is stored up in a latent form; that is, it is used, not as heat, but as force, and is employed in bringing the atoms of water into new relationship; during the night the vapor returns to its former state as water, and the heat-force again becomes sensible heat. Thus the day is cooled and the night made warmer.

"Ansel, have you ever heard the 'dew point' spoken of?"

"Yes, sir, I have."

"Do you know what is meant by it?"

"That point or degree of temperature at which dew begins to be formed."

"Upon what does the dew point depend?"

"Upon the amount of vapor in the air."

"That is right, Ansel. If at any time the full possible amount of vapor should exist, any diminution of the temperature must, of course, cause dew to be deposited. Do you know, Ansel, how to ascertain the dew point at any time?"

"No, sir, I do not."

"There is a beautiful instrument known as Daniell's Hygrometer which shows the dew point as a thermometer shows the temperature. But any one can easily determine the dew point without a special instrument for that purpose. Pour warm water into a glass pitcher or goblet whose outer surface has been wiped perfectly dry, and polished. Into this set a common thermometer. Cool down this warm water by dropping into it small pieces of ice, and notice carefully when the polished glass begins to be dimmed as if it had been breathed upon. When that begins to take place the thermometer will show the dew point. In this manner we can determine the amount of vapor in the air, and by estimating the probable temperature of the night judge of the probability that dew will fall."

"I have noticed some things," said Peter, "about the formation of dew which I do not understand, and I wish very much to ask about them."

CURIOSITIES OF HEAT.   185

"I should be glad to hear your questions, and will answer them if I can."

"I have noticed that dew falls on clear nights, but not very often on cloudy nights. I don't see why that is so."

"Have you ever noticed whether cloudy nights or clear nights are the warmer?"

"Cloudy nights are commonly warmer, I think, but I never could see the reason for that, either."

"Can you tell why a newspaper spread over a tomato vine keeps the frost from the vine?"

"Because the frost comes upon the paper instead of the vine, of course."

"But why do you say, of course? Why does not the dew—for frost is nothing but dew frozen as it forms—come upon the under side of the paper?"

"How could the dew fall upon the under side?"

"That is just the point which we need first of all to understand. Men commonly speak of dew as if it fell. I don't know but I have spoken of the falling of dew in this lesson. But dew does not fall at all. The vapor simply touches some cold object, and is condensed upon it. The vapor by its elasticity presses against the cold body, and the process of condensation continues until either the body is warmed by the heat set free so that its temperature rises above the dew point, or till the vapor is so far exhausted that the dew point falls below the existing temperature. Dew is formed upon the upper surface and not upon the under, because the upper surface is cool and the under surface is warmer. Beneath the paper spread over the tomato vine, the earth is radiating heat and the paper is radiating it back again. If the paper were not there, the heat would be radiated into space and not returned again. The vine would soon radiate away its little store of heat, its temperature would sink below the dew point, and dew or frost would be deposited upon it. The under surfaces of objects are kept warm by the radiation from the earth. In the same manner clouds are wrapped around the earth and keep it warm by radiating back its radiant heat. Dew is not formed on cloudy nights, because they are warmer: the clouds throw back the heat which otherwise would be lost in open space."

"I never knew before," said Peter, "that

clouds were of any great use except to send down rain."

"We shall see in the course of our lessons that clouds are of very great use in warming the earth in other ways, as well as by serving as blankets and radiating back the heat which otherwise would escape."

"I wanted also to ask why dew falls—I mean, is formed—on grass and leaves of plants while stones are dry."

"I will answer your question by asking another. Did you ever see barefoot boys running in the cold dew stop and stand upon a stone or rock to get their feet warm?"

"Oh yes, sir; I have done it myself."

"Why did you stand upon a rock?"

"Because I had learned that the rocks would be warm."

"I think that answers your question. The rocks and stones are warmer than the grass and the leaves. The blades of grass and the leaves are thin and pointed and rough, and have a very large radiating surface. They have but little heat, and that little they part with rapidly. The rocks and stones, on the other hand, are bulky, and contain a much larger store of heat, their radiating surface is comparatively small; only one side is exposed, the other being covered by the warm earth, from which they are drinking in heat almost as rapidly as they lose it. They therefore do not lose heat enough to sink their temperature to the dew point.

"So much, then, for the means employed to moderate the changes of temperature from day to night and from night to day. But upon the sea-coast and upon certain islands of the sea another agency is employed. Will some one suggest what this agency is?"

No one else answered, and finally Mr. Hume said: "I suppose, of course, that you refer to the land and sea breezes?"

"This is what I had in mind. During the day the land is warmer than the sea, and the breeze from the sea blowing upon the land cools the air. During the night the land radiates its heat more rapidly than the water, and soon the sea becomes the warmer. Then a breeze springs up in the opposite direction; the cooler air of the land flows out upon the sea. By this means the air upon the land and the air upon the sea are daily commingled, thus securing a more even temperature upon the land. This softens the extremes of daily temperature. I make only this brief reference to the land and sea breezes, because in another connection we shall examine the general subject of winds and their influence in the equalization of temperature.

"The result of all these influences is that the changes of temperature from day to night and from night to day, while not inconsiderable, are by no means destructive, and in many cases are no greater than is refreshing and agreeable. These agencies remind us every day of the wise provision of the Creator for the well-being of his creatures. 'Day unto day uttereth speech and night unto night showeth knowledge. There is no speech nor language where their voice is not heard.' This care for the earthly well-being of men is but a type of his care for their spiritual happiness. The plan of salvation, and the ways of divine providence working in accordance therewith, are more wonderful both in their means and their end than the greatest of the works of Nature. If while we study the natural we forget the supernatural, we commit the greatest mistake: we pass by the greater to examine the less. The natural is valuable only as it leads to the spiritual."

CHAPTER IX.

THE MINISTRY OF SUFFERING.

"YOU must know, Mr. Wilton," said Mr. Hume, "that my mind is full of objections, whether I speak them out or keep silence. I have looked so long upon one side only that I find it hard to look upon other sides also; and if there be a Satan, as the Bible teaches, I think he must be marshaling all his legions to overwhelm me by the force of his impetuous assaults. I cannot disguise the fact—I do not attempt to disguise it—that my mind is not at ease. It used to be at rest, at least comparatively so—not happy, yet not agitated and distressed. My heart was not satisfied, but I believed that my position of unbelief was logically impregnable. But I confess it, my unbelief has of late been shaken. I am no longer contented. How I came into this state, I do not know. I am certain that my present unrest was not produced by the force of arguments which I had heard. It seems to me as if it sprang up uncaused. The old arguments which I have thought impregnable do not now satisfy me. Why, I cannot tell. I think this statement is due to you to explain my position in your Bible class, and also to prepare the way for a question which I wish very much to propose. I have no objections to make to the marks of wisdom and benevolent design seen in the works of creation which I cannot myself answer and remove. Good-will and goodness to the inhabitants of the earth lie on the very surface of things; or, if I go beneath the surface, I find them no less manifest in the profoundest and subtlest arrangements of the universe. If I say, 'This is all the work of chance,' my very language is self-contradictory and looks me out of countenance, for the very idea of chance is the opposite of wise and orderly arrangement. The difference between design and chance is that the one works by orderly arrangements adapted to the accomplishment of a foreseen end, while the other shows itself in chaotic disorder, with no adaptation to the accomplishment of a purpose. To say that a universe like this, filled in every part with order and beauty, with subtle and unseen elements and agencies working out into the boldest relief in the accomplishment of beneficent ends, all minute elements blending in the sublime sweep of the universal plan,—to say that such a universe is the work of chance is to use language without meaning.

"If I deny a providential plan in the creation and government of the world, and attribute to brute matter a nature that, by its own inherent force, spontaneously develops into all these contrivances of use and beauty, I see that the wisdom of the whole universe is concentrated in the nature of matter, and, if it be possible, infinite subtlety of design is doubly manifest. To create a machine which, upon its elements being thrown into an indiscriminate pile, shall arrange itself, adapt part to part, and set itself in motion; which shall repair all its breaks, produce other machines as curious as itself, and thus reproduce itself and perpetuate its existence for ever—that would certainly be the acme of intelligent design.

"Or if I go farther and deny a Creator, ascribing to the universe an eternal, uncreated existence, I see that I only entangle myself in a complication of difficulties. I find myself standing face to face with the best-established facts of geology. If the fact that the animal tribes which inhabit the earth, and especially the human race, had a beginning be not well established, then no fact in geological science can be reckoned as fixed. Geology has overturned the idea of an infinite series of generations of animals and men. Nor do I see that I gain any advantage or give any explanation of the universe by attributing to matter everything which others refer to an intelligent and almighty Creator. The distinction between mind and matter is that mind is endowed with intelligence and will, while matter has neither intelligence nor will, but only blind forces, blind attractions and repulsions. If I attribute the order, beauty, design, and benevolence of the universe to mere matter, I clothe matter with the attributes of spirit. In fact, I only set up another God and ascribe to the universe a true divinity. I make myself a kind of pantheist, investing all matter with the attributes of mind and spirit. All this I have pondered over for many a day, and I cannot deny that a belief in an intelligent Creator of the universe is logically more satisfactory. But there is one question which confronts me at every turn. I suppose that I might at length work out an answer for myself and that I should now see the explanation if all my thinking for so many years had not been upon the other side."

"I am afraid that I shall not be able to give you satisfaction," replied Mr. Wilton, "but I shall be glad to hear your question. I can at least appreciate your state, and sympathize with you in your groping and struggling. I am glad that you are walking the road you have just described. You say that you do not know what has brought you to your present state. I can easily tell you: your experience at this point is not singular; I think the Holy Spirit of God has been leading you and has brought you to your present position. I trust in God that he will lead you still farther. You have great cause for thankfulness and great cause for trembling. Let me caution you: be careful how you treat the divine Spirit; walk softly; be honest, sincere, and simple-hearted as a little child. 'Except a man become as a little child, he cannot see the kingdom of God.' Above all things, be sincere and straightforward. Deal truly and frankly with the Spirit. If you will only be honest and frank,—honest and frank to yourself, honest and frank to all men, honest and frank with God,—God will soon give you cause to praise him and love him for ever and ever. But what is the question which you wished to propose?"

"My difficulty is this: Along with the many arrangements for conferring enjoyment and promoting the well-being of man are other arrangements for suffering. Man is made as capable of suffering as of enjoyment, and there are appliances provided which are certain to inflict that pain of which man is capable. How is this provision for suffering in man and in all sentient creatures consistent with the benevolence elsewhere shown? How are we to combine these two sets of arrangements in our thinking?"

"A full unfolding of the ministry of pain in the good providence of God would lead us entirely aside from our course of study."

"But for me," said Mr. Hume, earnestly, "it would be not at all aside; for if I can once see that the provision for suffering made in the constitution of man and of Nature is not repugnant to the idea of a wise and good Creator and Disposer of human affairs, I will admit whatever you shall have to say afterward, and I shall feel that the gospel of Christ comes to man and comes to me with a moral force which ought not to be resisted. I know that I have no right to come into your class and ask you to turn aside from your course of study, and the gospel certainly owes nothing to me, yet I do hope you will give the opinions which you hold upon this subject, if you have formed any positive opinions."

"I am sure," exclaimed Peter, "that we shall all be very glad to have you spend the time of this lesson in speaking of this subject."

"But how would it please you if my talk upon the ministry of pain should prove to be very much like a sermon?"

"I think we like your sermons. I know that we were never so much interested in them as now."

"Very well, then; I will give you, as Mr. Hume says, some of my conclusions touching this matter of pain and suffering; and if my opinions are not satisfactory or do not cover the facts in the case, it will not be because I have given the subject little thought or have had little experience of suffering. The Lord has led me by a rugged road; he has given me tears to drink and mingled my cup with weeping. But for this I thank him, and I expect, when I shall look back from the life to come upon my earthly course, to see my days of pain and grief shining more brightly than the hours of radiant sunshine.

"First of all, then, I believe that with the clear exhibition of benevolent design which we see in this world we ought not to doubt the goodness of the Creator, even if we can give no rational explanation of the suffering which abounds. We ought not to believe, we cannot believe, that the Creator's own attributes are self-contradictory and antagonistic, that the same infinite Being is both good and evil, partly benevolent and partly malignant. If God is good at all, he is wholly good. Nor can we believe that a good being and an evil being—God and Satan—hold joint sway over the universe and co-operated in the work of creation, and that the good is to be ascribed to the one, and the pain and suffering to the other.

Whether we can explain it or not, we must believe that there is a good reason for the existence of suffering; unless, indeed, we count the infliction of pain the chief end of the creation, and refer the happiness which men enjoy to some incidental arrangements not contemplated as important in the work of creation. But no sane man can think that this world is the work of a demon seeking to fill the earth with groans and wretchedness. Our consciences and our reason alike require us to believe in the supremacy of goodness.

"In presenting my views, I of course cannot attempt to prove everything from the beginning: I must take some things for granted between us. We must start with the admission that there is a God, and that he is a righteous, moral governor. We must at least believe what Paul declares to be needful: 'He that cometh to God must believe that he is, and that he is a rewarder of them that diligently seek him.' We must also believe our own consciences when they testify that men are responsible, free moral actors, and that sin and guilt are not false notions arising from diseased and morbid mental conditions, but realities, true ideas which arise in the mind when it works as

God designed. Do you freely admit these points of belief, Mr. Hume?"

"Yes, sir; I could not ask you to prove every point touched upon in the argument, for that would require half a score of volumes, nor will I deny the testimony of my own conscience that there is a God, and that men are rightly responsible to him."

"Starting, then, with these fundamental principles, we will look first at the provision made for physical pain. Men and, I suppose, all living creatures are created with the capacity of suffering. The same nerves of sensation which if excited naturally give rise to pleasure may be excited unnaturally and inflict pain. But why not endow living creatures with nerves of sensation which could experience pleasure, but could not feel pain? Is this possible? Perhaps so, but no man can affirm it with certainty. I do not think that any man can clearly conceive such a thing. To us the capacity of enjoying and that of suffering seem inseparable. But there is no need of insisting upon this point, for the capacity of feeling pain is a most benevolent provision of the Creator for the benefit of living creatures. It is designed to save life and limb.

Pain is the sentinel set to guard the outposts of the citadel of life. If there were no pain, men would thrust their hands into flames without knowing it. They would indulge in all manner of destructive excesses, and no sufferings would warn them of danger. They would drink poison, and no pain would bid them make haste to take the antidote. Tear men limb from limb, hew them in pieces with the sword, and no painful sensations would rouse them to self-defence. Without this benevolent provision of pain the race of man could hardly be saved from extinction. How much more would this be true of the animal tribes, which are wholly dependent on instinct for guidance and impulse to action! We accordingly find pain possible in those parts of the body where pain can subserve the purpose of protection; elsewhere no provision is made for pain. Nerves of sensation abound in those parts which require especial care or are especially exposed. The skin is exposed, therefore the skin is well supplied with nerves. The parts beneath the skin are less exposed, and are injured only by first wounding the skin; they are therefore less sensitive. The heart, though so very important, is almost insensible to pain, because the capacity of suffering at that point would confer no protection. The eye is delicate and requires the greatest care, and to secure that needed care the Creator has made it delicately susceptible of pain. The sole of the foot, as its work demanded, was made capable of bearing the roughest usage, and hence the sole of the foot is but little supplied with nerves of sensation. Still farther, when on account of injury any part of the body requires unwonted care, provision is made that the injured part shall become especially sensitive. A bone when well and sound may be cut or sawed almost without pain, but when the bone is injured it becomes inflamed and feels pain most keenly. When a limb for the sake of its own safety ought to be kept quiet, Nature makes it painful to move it. For the benevolent object of preserving life and guarding the well-being of living creatures pain is given. The provision for pain shows the presence of danger, the liability of receiving injury, and the kind design of putting men on their guard. It is the automatic guardian of our happiness. This is all that I have to say about bodily pain.

"Mental suffering and pain of conscience are

designed, first of all, to subserve the same purpose. The sense of guilt when a man commits a wicked act is designed, first, to lead him to repentance. It is the divine alarum placed within the soul to remind men that they have done evil and received moral damage which must be repaired. It is the moral goad which pricks men to warn them to turn from wickedness. If evil-doing were as pleasant as well-doing, men would see no difference between right and wrong; all moral ideas would be subverted and the glory and beauty of man would be trailed in the dust.

" But a guilty conscience continues to trouble wicked men after the day of repentance has passed; Remorse indeed seems to rise up with preternatural power when Mercy has withdrawn for ever from the sight of Hope. What is the meaning of this? It means that which we admitted in the beginning, that sinners are guilty in God's sight, that guilt is a real thing and deserves punishment, and that God, the holy and righteous King of men, does actually punish the guilty. God is holy and abhors sin. Remorse of conscience is the shadow of the Creator's frown, the voice of his eternal indignation echoing and re-echoing in the soul of man. It is the divine wrath penetrating the human spirit and making itself felt. As the holy God abhors sin for ever, the wicked must expect to feel that abhorrence for ever. He who puts himself into a rebellious position toward his Creator must stand in that unnatural attitude guilty and suffering. We can conceive that this should be otherwise only by subverting the foundations of the moral world. Beings created in the image of God, created with a conscience and moral affections, created with moral freedom, can attain blessedness only by aspiring to heavenly things and becoming God-like. If they break away from the divine will and order, they must suffer the divine frown, they must feel that frown. How can God make his frown felt except by looking pain, so to speak, into the sinner's conscience?

"But this whole subject of pain and suffering derives a double significance from the fact that the human race is a fallen race, alienated from God by wicked works, yet under a merciful dispensation in which they are called to return to obedience. There is no moral quality good and beautiful to our eyes or pleasing to God in which men are not altogether lacking, and what is still worse, men grow in evil; their last state is worse than the first. There is no healing power in the man which can renovate his heart and bring him back to holiness. It would seem as if some satanic power were hurrying the human race along the road to ruin. If men are to be saved, it must be by a force of renovation outside of themselves, which shall reverse the evil bias of their nature. You say that the world seems fitted to develop man's capacity for suffering, and that this appears to be as much a part of the divine plan as the impartation of happiness. What, think you, would be the result if the human race were planted in a world where nothing could give pain, where everything would afford gratification? What, Mr. Hume, do you think the effect would be upon creatures such as we all know men to be?"

"I hardly dare answer with the little thought I have given to the subject. I would rather listen than speak."

"I have noticed," exclaimed Ansel, "that those boys who have everything done to suit them at home are the most unmanageable in school and the most disagreeable to play with."

"Picture to yourselves," continued Mr. Wilton, "a man who from childhood should have nothing to suffer, no pain or weariness or hardness to bear. From childhood he has no bodily pain, and the comforts of life are so carefully and bountifully provided that he receives no unpleasant sensation. Winter never chills him, summer never heats him. His slightest wants are all anticipated. All his sensations are pleasure. Let the same be true of his mind. His will is never crossed; whatever he wishes is given him; there is no call for self-denial or self-control or abstinence or patience. He feels no pressure of need spurring him to exertion. His whole life is enjoyment. His very body would grow up, not strengthened and compacted for exertion, but fitted only for the softness of indolence and ease. His will would be the selfishness of self-will rather than an intelligent, reasonable self-control. There would be no tenderness and power of love, no endurance and patience in labor, no strength of moral purpose under temptation, no self-denial and self-sacrifice of love for the good of others or for the attainment of a higher blessedness, no faith in God nurtured in darkness and trial. We should have a mushroom growth of luxurious tastes and indolent ease, impulsiveness and impatience, strength only in selfish, passionate self-will and rampant, luxuriant vices. No other result would be possible with creatures like us. Strength is developed only under circumstances which call for the exercise of strength. A certain hardness and hardihood of living is needed to develop a manly body. Resolute intellectual exertion in the face of difficulties is demanded to educate the mental faculties. An earthly life not wholly satisfactory is needed to awaken in faithless men a longing for a better land. We may look upon the sufferings of this world, taken as a whole, as an expression of God's displeasure at sin. How very much is such an expression needed! If life were nothing but pleasure, how completely men would forget sin and duty and God and heaven! All the varied experiences of joy and sorrow, of good and ill, of trial and triumph, are needed for man's spiritual discipline. I think you will bear me witness that the noblest, sweetest, most beautiful characters are found in those who have drunk the cup of sorrow to the dregs."

"I cannot deny it, Mr. Wilton. There is old Deacon Smith. We all know something of his history, I suppose. He was a poor boy; when he was twelve years of age his father died, and his mother died four years later. But he worked his way, first to a good education, and then to an honorable position and ample fortune. Then the dishonesty of a partner brought him back to poverty too late in life for him to recover himself. Now in his old age he works for a small salary in the office of another. But he is as cheerful and as grateful as if he had all that heart could wish, and had never in his life suffered a pang. I think he verily believes that everything which has befallen him has been an expression of God's love for him. He sheds no tears except for the griefs of others. I think he truly rejoices with those that rejoice and weeps with those that weep. As for faith in God, I suppose he would go into a lion's den as calmly as did Daniel. If every professor of religion were like him, I am sure that nobody could say a word against the gospel. I freely confess that Deacon Smith's character has affected me more than all the arguments I have heard in favor of Christianity."

"As to that, Mr. Hume," replied Mr. Wilton, "we have both of us, doubtless, seen men who would hate a man the more bitterly in proportion as he should show himself Christlike. And as to every church-member being like Deacon Smith, we could hardly expect such a character to be nurtured in a day or a year. Deacon Smith has become what he is by a lifetime of severest spiritual discipline and patient endeavor. Such characters are wrought out only by a discipline of every form of trial. This world is constituted as it is for the purpose of giving just such a discipline of effort and patience.

"This explanation brings us, however, only to the vestibule of the great mystery of suffering in the work of recovering man from the Fall. The Captain of our salvation, who put himself in man's place and took upon himself all human conditions, was made perfect through suffering. The full preparation for his work as the Saviour of man called for a discipline of pain. I shall not attempt to explain this experience of Christ, but salvation brings the believer into a state of profoundest and most mysterious union with

Christ. The believer must walk in the footsteps of Jesus. As Christ first came into a condition of sympathy with man, so must man come into a condition of sympathy with him. The believer must share and repeat, in a feebler way, of course, the experiences of the Lord Jesus. He must fill up that which is behind of the sufferings of his Saviour. By this union with Christ in the discipline of pain the Christian is prepared for a union of blessedness. 'If we suffer, we shall also reign with him.' How broad and deep this union of the believer with Christ may be, I cannot tell. I am not able to measure this idea. It seems to me like one of God's infinite thoughts, revealed in its dimness to overawe the souls of men by its shadowy sublimity—seen only enough to suggest how much vaster is that which remains unseen—an iceberg, one part standing out and nine parts sunk in the unfathomed sea. It is a thought to be felt and experienced rather than weighed and measured by human logic. This is all that I have to say upon this subject. Do these views commend themselves to you, Mr. Hume?"

"I do not know," was the reply; "I want to revolve the subject in my own mind. I have received some new ideas, but I judge that a man needs experience in this matter as well as thinking. If I had Deacon Smith's experience of life, I could form a better opinion. As much as this I can see to be true—that provision for bodily pain is a safeguard to the happiness and life of men, and that a world which should anticipate every human want, leaving nothing to be struggled after and nothing to be endured, would have a disastrous influence upon human character. I will admit that the provision for pain is wise and good."

"One other point," continued Mr. Wilton, "we ought to notice before leaving this subject. The word of God says, 'We know that all things work together for good to them which love God,' but it says no such thing of those who do not love him. The afflictions of this life work out for the righteous 'a far more exceeding and eternal weight of glory.' The ministry of pain is a ministry of love only to those who submit to Christ. To those who kick at God's mercies the best blessings turn to evils and curses; to the faithful in Christ the greatest griefs and calamities become choice blessings. A submissive heart and the agency of the Holy Spirit are needed to sanctify pain. It is a great mistake to think that all men are made better by afflictions. Only the few get good from the discipline of life. With many persons troubles only stir up the worst passions till they rage like caged tigers."

"This last remark, Mr. Wilton, has thrown a flood of light upon this subject. But it seems strange to me to find myself saying this. I see how it is that so large a part of the pains of life is found in the end to accomplish no good. The evil remains evil. Do you think that my long trial of doubt and unrest and pain of heart can ever be blessed to my good?'

"That it can be so blessed to your good and to the good of many others I have no doubt; but whether it will be, I cannot tell. That depends upon yourself, upon your coming through Christ to God as your heavenly Father. It is my earnest prayer that from your unrest of spirit deep peace in Christ may break forth; and many others unite in the same."

"I certainly hope," said Mr. Hume, "that my life may not come to nothing. It seems as if something better than a few years of mingled pain and pleasure, overshadowed by most painful doubt and darkness and followed by a plunge into nothingness, must be possible for me."

"God give you grace," said Mr. Wilton, earnestly, "to forget the things which are behind, and reach out your hands toward the worthiest destiny! But remember that there is a destiny more terrible than to cease to be, there is a death deeper and darker than the grave.

> 'There is a death whose pang
> Outlasts the fleeting breath;
> Oh, what eternal terrors hang
> Around the second death!'"

Mr. Wilton did not think it best to attempt to draw out Mr. Hume farther at that time. He saw that he appeared to be under the guidance of the Holy Spirit, and hoped that he would soon experience the new birth by which old things pass away and all things become new. He knew that time is an element even in the operations of the Spirit, and he feared to shake the bough too roughly lest the fruit should fall untimely only to wither in his hand. Happily, the superintendent's bell brought the conversation at that point to a natural conclusion.

CHAPTER X.

TRANSPORTATION OF HEAT.

TO-DAY we come to that subject which we should have looked at a week ago, if that I hope not unprofitable discussion of the uses of trials and the ministry of pain had not prevented. We must now examine the arrangement for softening the rigors of winter and toning down the heat of summer. The general principle is that in summer the earth receives an excess of heat, while in winter the opposite is true. These extremes are mitigated by transferring heat from summer to winter. How is this accomplished? Any one who has thoughts upon this subject may answer."

"I have some thoughts," said Ansel, "but whether right or wrong, I cannot tell. I should think heat might be carried from summer to winter in the same way as from day to night."

"What are some of those means for transferring heat which seem to you to operate the same in the annual as in the daily changes of temperature?"

"One is the absorption and radiation of heat, and another is the evaporation of water and the condensation of vapor."

"You are right," said Mr. Wilton. "The effect of these operations in the equalization of the annual extremes of heat is in no wise different from their effect upon the temperature of day and night, but from summer to winter their effect is vaster and more impressive. During the summer, sea and land, and 'all that in them is,' are receiving heat and rising in temperature. The heat of summer penetrates and warms the earth nearly a hundred feet in depth. Into the sea heat penetrates still deeper. How vast the amount of heat required to warm the whole surface of the earth and sea to such depths! By withdrawing so much heat from active use the intensity of the summer temperature is softened. During the colder months the land and sea slowly radiate their heat. We can hardly over-estimate the effect of this alternate absorption and radiation of heat. So great is the effect of this stored up heat that the sea and the great lakes never freeze even in the coldest winter weather, except in the polar regions, and the temperature must fall far below freezing and continue for a long time below the freezing point before the earth begins to freeze. The great bodies of water, remaining always at a temperature above thirty-two degrees, are especially important in warming the wintry air. In the coldest weather they seem like steaming caldrons throwing up their warm vapor. It is the absorption and radiation of heat alone which prevent the temperature of the atmosphere from rising or falling suddenly to the highest or lowest point possible. The sun breaks forth in all its splendor at noonday in summer: what if the sun were to remain stationary, shining thus in his strength for days and months? Everything would be consumed with heat. But why do not the glowing rays of the sun raise the temperature at once to the highest possible point? Because the earth and sea and every object upon the earth absorb the heat, storing it up and holding it in reserve. On the other hand, when the sun sets and his heat is withdrawn, why does not the temperature fall suddenly to the lowest possible point? Because the heat held in store is slowly radiated and the change of temperature rendered gradual.

"In this work of absorbing and radiating heat every object, earth, air, and sea, does its appropriate share. But water stands chief, and performs the largest service. Its high specific heat enables it to hold in store the largest calorific treasure, and causes it to change its temperature more slowly.

"The formation and condensation of vapor also operate in the same manner as in the transitions of day and night. During the summer the higher average temperature makes it possible for a much larger amount of vapor to be formed than in winter. You remember that at eighty degrees vapor equal to thirteen inches of water can sustain itself, while at thirty-six degrees the elastic force of vapor is equal to the pressure of only two inches and two-fifths of water, and at four degrees to three-fifths of an inch. If the mean summer temperature at any place were eighty degrees, it would be possible for more than one foot of water to be held in the form of vapor. In the formation of this vapor heat would be consumed sufficient to boil more than five and a half feet of ice water. If the mean winter temperature at the same place be thirty-six degrees, more than three-fourths of this vapor must be condensed and give out its latent heat to warm the air. It is not to be supposed that the full amount of vapor which can support itself does commonly exist, but the difference between the average amount of vapor in summer and in winter must be very great. I suppose this difference often amounts to four or six inches of water. If we suppose it to be four inches, an amount of heat is transferred from summer to winter sufficient to boil twenty-two inches of ice water. In estimating the effect of this we must consider that this heat is not given out gradually and regularly for three months, but whenever there is a sudden fall of temperature vapor is condensed, latent heat becomes sensible, and the suddenness and intensity of the fall are diminished. We need also to bear in mind that every open body of water is sending up its clouds of vapor constantly. The open lakes, and especially the sea, are like a seething caldron; and thus immensely more vapor is condensed during the winter months than is brought over from summer to winter. Much of the vapor formed in winter is to be set to the account of summer, for it is the summer's heat absorbed by the water, which maintains its temperature and enables it to throw up such clouds of vapor, even in midwinter. But this comes in more properly at another place, and we will leave it for the present.

"There is another transition experienced by water by which heat treasured up in summer is made available for softening the rigors of winter. Who will suggest it?"

"It is the freezing of water," said Mr. Hume. "In the process of crystallization one hundred and forty degrees of latent heat become sensible."

"And this," continued Mr. Wilton, "is no inconsiderable matter. Every pound of water frozen upon the surface of our lakes and rivers, every pound of water frozen in the wet earth, every pound of water frozen as snow or sleet in the air, gives out as much heat as would boil an equal amount of water at seventy-two degrees. Have you never heard of setting tubs of water in cellars to keep vegetables from freezing?"

"I have," replied Peter. "I visited my grandfather two years ago, and his cellar sometimes froze. I asked him why he put tubs of water in his cellar, but he could not tell me, only he said that he knew that tubs of water in his cellar did keep his vegetables from being nipped with the frost."

"Can you tell us, Peter, why tubs of water set in a cellar should have this effect?"

"I suppose that when the water begins to freeze it begins to give out its latent heat."

"That is one part of the reason. The water is drawn from the well at perhaps fifty degrees; it must lose eighteen degrees of heat before it begins to freeze, and all the heat which the water loses the air of the cellar gains. And then, as you said, as soon as the water begins to freeze latent heat begins to become sensible. Every pound of water frozen sets free heat enough to raise a pound of water through one hundred and forty degrees. But why do not the vegetables begin to freeze as soon as the water?"

"I don't know."

"Water holding salt or other minerals in solution freezes at a lower temperature than pure water. For this reason the juices of vegetables and fruits and the sap of trees may be cooled below thirty-two degrees without freezing. On this account the water set in cellars tends to prevent vegetables from freezing; the water begins to freeze at thirty-two degrees, while potatoes and turnips may be cooled a little lower than thirty-two degrees without harm. In this manner the buds of trees are sometimes warmed and protected by the coating of ice which forms around them. The drops of water, falling through the sleety air, touch upon the twigs of trees and freeze upon them, an icy coat embracing them all around. In freezing, the water gives out one hundred and forty degrees of heat, a part of which goes to the air and a part to the twig."

"This reminds me," said Ansel, "of what the Irishman said on being told that snow contains heat, that 'it would be a blessed thing for the poor if one could tell how many snowballs it would take to boil a tea-kettle.'"

"It might be difficult to use snowballs to boil the tea-kettle, but the heat given out in the formation of the snowflakes is doubtless employed quite as usefully for the poor as if used in preparing their tea. You have all noticed that before a snow-storm, or perhaps during the early part of the storm, the temperature generally becomes milder, and you have often heard the remark, 'It is too cold to snow.' Men have learned that the coming of a snow-storm is attended by a warming of the air. This popular impression is philosophical, yet few understand its philosophy. A foot of snow falls, equal to two or three inches of water. In the condensation of the vapor which formed this snow one thousand degrees of latent heat become sensible, and then in the congelation of the clouds into snowflakes one hundred and forty degrees of heat are evolved. This softening of the rigors of winter is, I think, as great a blessing to the poor as the heating of the tea-kettle. Let us make an estimate of the amount of heat set free in the production of one great snow-storm. Two feet of snow falls, equal, we will suppose, to five inches of water. In the condensation of the watery vapor one thousand degrees of heat are evolved, and in the congelation one hundred and forty degrees—an amount of heat which would boil three feet of cold spring water. In every square mile there are 27,878,400 square feet, and a square mile of water three feet in depth would contain 83,625,200 cubic feet. The production of such a snow-storm sets free for every square mile of surface heat which would boil more than 80,000,000 of cubic feet of spring water. Such a storm sometimes extends over a region of country a thousand miles square, that is, over a million of square miles. In the production of one such storm—a very heavy and extensive storm, I have supposed—heat is generated which would boil eighty millions of millions (80,000,000,000,000) of cubic feet of spring water—an amount altogether too vast for our comprehension. To accomplish this result by combustion would require more than 500,000,000 of tons of anthracite coal—an amount at least three times as great as the yearly product of all the coal-mines of the world. And this is but one heavy storm. The amount of rainfall in the United States may be thirty-six inches or forty or forty-five inches. Supposing the average rainfall of the whole earth to be twenty-four inches—an estimate very far below the truth—we have this result: There are, in round numbers, two hundred millions of square miles of surface, more than five and a half quadrillions (5,575,680,000,000,000) of square feet and more than eleven quadrillions of cubic feet of water.

The condensation of this amount of vapor would boil more than sixty quadrillions of cubic feet of ice water. One pound of anthracite coal burned under the most favorable circumstances will boil sixty pounds of ice water. To boil sixty quadrillions of cubic feet of ice water would require sixty quadrillions of pounds of coal—thirty billions of tons—not less than twenty-five tons to every inhabitant of the globe. At this rate a very few years would exhaust the coal-fields of the world. Calculations like these are useful in showing upon how stupendous a scale the Creator carries on his operations. But we must remember that these works are carried on, not to amaze men, but to benefit them. The works go on silently and unseen, challenging no attention from fools, receiving no thought except from the patient student of Nature, and eliciting no thankful recognition save from a few reverent worshipers.

"But I have been led away from a point which I had in mind. While considering the effect of heat in expanding bodies, I reminded you that water presents a marked peculiarity, and promised to speak of it more fully. This is the place for us to look at this singular and beautiful peculiarity of water. What is the general principle touching the effect of heat upon bodies?"

"Heat expands bodies and cold contracts them," answered Ansel.

"Water both illustrates this rule and presents some very interesting apparent exceptions. It contracts by cold like other bodies till it reaches the temperature of thirty-nine and a half degrees; it then begins to expand, and expands regularly till it falls to thirty-two degrees; at that point it freezes, and in freezing it expands at once about one-ninth of its bulk. If the cooling process be continued, the ice produced contracts like any other solid. This peculiarity of the interrupted and unequal expansion of water is of the utmost importance in the affairs of our world. Consider the result if the water were to contract by cold as do other bodies down to the freezing point and below it. Water is cooled from the top by contact with the cold air. As the upper film of water cooled it would sink and a new stratum be brought to the surface; that in turn would be cooled and sink, and thus the cooling process would go on with the utmost rapidity till the whole body of water should be reduced to the freezing temperature. Then congelation would begin, and the first particles of ice formed would sink to the bottom, and as fast as the water became frozen at the top the ice would sink. In this manner a solid body of ice would be formed at the bottom of our lakes and rivers, while the surface would remain unfrozen in contact with the cold air till the whole body of water became a compact mass of ice. Great lakes turned to solid ice would not be thawed during the whole of the summer, for the water warmed from the top would not sink, but would form a warm stratum of water upon the surface, while, below, the solid ice would lie hardly feeling the summer heat. Nay, more; in the higher latitudes it would seem as if the very ocean must be turned to solid ice, never to be melted till the end of time. By the singular expansion of water below thirty-nine and half degrees and its great expansion in congelation, these disastrous consequences are prevented. Our lakes are cooled even in winter only to thirty-nine and a half degrees; below this temperature the colder water is lighter and remains upon the surface; ice floats upon the surface. The top becomes ice, but the great mass of the water remains at thirty-nine

P

degrees, and the inhabitants of the waters live on unharmed. Spring comes, and the ice, being upon the surface, is soon melted, and the unbound waves begin again to ripple forth their unconscious joy."

"Do you look upon this irregular expansion and contraction of water," asked Mr. Hume, "as a real exception to the rule that heat expands bodies?"

"Not at all. In freezing, a new force comes in and asserts itself—the force of crystallization; or, more exactly, as the force of heat diminishes the force of crystallization becomes predominant, and throws the atoms into new positions and new relationships. To this new arrangement of atoms is due the expansion in freezing. Ice contracts and expands by cold and heat the same as any other solid. The attraction of crystallization begins, doubtless, to throw the atoms into their new and crystalline arrangement at the temperature of thirty-nine and a half degrees.

"We must remember that the heat which is set free in the condensation of vapor and in the freezing of water is absorbed in the formation of vapor and the melting of water. As much heat is taken from summer as is conferred upon winter. The summer is cooled as much as winter is warmed. The formation of vapor is a cooling process. Water is prevented from rising above the boiling point by the formation of vapor. Perspiration cools us by the evaporation to which it gives rise from the whole surface of our bodies. And the higher the temperature, the more rapid the evaporation, and the more vigorous the cooling process.

"We might look at other appliances for transferring heat from summer to winter, but they belong in principle to another department. We have now looked at some of the means for transferring heat in time. The heat is treasured up at the heated noonday, to be brought out for use during the cool hours of night; it is garnered from the excessive heats of summer to supply the deficiencies of winter. It is laid up in store to-day to be expended at any future time when needed. The transfer is a transfer not in space, but in time. We must hereafter examine those arrangements by which heat is transported through space. Some of these arrangements exert an influence upon day and night and upon summer and winter, and thus throw further light upon the subjects already discussed. Already more than once topics have been suggested and their full consideration put off till some more fitting time. In our next lesson we must begin the examination of these new principles. We have before spoken of the vicissitudes of days and seasons and years. We shall now have to do with the vicissitudes of zones and lands and seas, of deserts and mountain ranges. The elements become vaster, the stage is broader, and the movements more sublime.

"I am glad that you are so well interested in these great and beautiful works of God's wisdom and power, but I hope that you do not forget that the crucified Christ is pre-eminently the power of God and the wisdom of God. These natural works are but the husk of which salvation from sin by Christ is the kernel. These outward things are wonderful and beautiful for the setting, but the gem, the royal precious stone, the Koh-i-noor, the 'mountain of light,' for which the setting was made, is the true knowledge of the true God and of his Son Jesus Christ. During the past few weeks you have heard others asking, 'What shall we do to be saved?' I should be greatly guilty if I allowed you to think earth, air, and sea, with all their silent and solemn movements, more important than our spiritual attitude toward God the Father and Christ the Saviour. Are you, Samuel, in your interest in studying Nature, forgetting Christ and the souls of men?"

"I hope not, and I think not. During the three years since my baptism I have never felt so much my obligation to Christ as now. I never felt before so deep a desire that my friends should repent and believe in Jesus. I think the love of Christ constrains me. I have not felt before that my work was very important; I have been expecting to work more earnestly by and by; but lately I have felt that Christ gives me something to do now for which he holds me responsible."

"What have you tried to do for Christ?"

"I have been praying for some of my young friends, and especially for Ansel and Peter. And then I felt that I must talk with them as well as pray for them."

"And can you, my young friends, be careless about your own salvation while Samuel is so anxious for you? Are you contented to live 'having no hope and without God in the world'? Is your happiness here and hereafter more important to Samuel than to yourselves?"

"We are interested," said Ansel. "We have been talking together about being Christians, but we don't know what to do."

"They said," broke in Samuel, "that they wished I would ask you to preach a sermon and tell them what they must do to be saved. They wished to go on with these lessons, but they thought that perhaps you would be willing to preach a sermon just upon that subject."

"You know that I often speak of that subject, and when persons have come to the inquiry-meeting I have told them what they must do. But I know that there must needs be 'line upon line.' If Ansel and Peter wish it, I will devote a sermon to the subject, and make it as plain as I can. Hardly anything gives me more pleasure than to explain the way of salvation when I know that my hearers are interested."

"We do wish to have you preach upon that subject, and I am sure that you will have a great many interested hearers besides Ansel and myself."

"But, Samuel, did you not pray for Mr. Hume also, and talk with him?"

"I prayed for him, but I was afraid to speak with him. I have tried to pray for him a double portion because I could not speak with him."

Tears gathered in Mr. Hume's eyes; the thought came to him that his unbelief had raised a barrier between himself and both God and his people. This pious young man was afraid to come to him lest he should meet the scornful arguments and cold derision of a proud unbeliever. He felt humbled—he, a subtle, well-read unbeliever, and Samuel a pious lad yearning for the salvation of his soul, but daring only to pray in secret for him.

"Have not you, Mr. Hume, been treating Christ and the Holy Spirit as Samuel feared that you would treat him?"

"Perhaps so," he answered. "I am sorry that Samuel did not come to me freely. I think he need not be afraid of me now. I also hope you will preach the sermon which Ansel and Peter wish to hear."

Mr. Wilton assured them that he would do as they wished unless the Spirit clearly drew him to some other subject. "I always look," he said, "to the Holy Spirit for direction in my preaching. 'When he, the Spirit of truth, is come,' said Jesus, 'he will lead you into all truth.' This was fulfilled pre-eminently, I suppose, in the inspired men who laid the foundation of the Church, but the Spirit still dwells in believers and leads those who love and follow Christ. The preacher of the gospel can do nothing without the power of the Spirit of God."

And I, kind reader, will give you the outline of the sermon if the Spirit bids him preach it.

CHAPTER XI.

AN EFFECTIVE SERMON.

MR. WILTON preached the sermon spoken of at the close of the last chapter the next Lord's Day morning. The more he thought upon the matter and inquired the mind of the Spirit, the more he felt that for a purpose the Spirit was calling him to unfold again the authority of God and the conditions of salvation. He gave notice of his subject, and invited all good men to pray that he might be able, like a good and wise steward of the mysteries of grace, to bring forth out of the treasure-house things new and old, and that the word might prove as a nail fastened in a sure place by the Master of assemblies. Much prayer was offered, and the people came together in a spirit of unwonted solemnity and earnestness.

Mr. Wilton prayed to the glorified Redeemer for his blessing: "O thou exalted Christ, we assemble in thy name and by thine authority. Thou hast bidden us not forsake the assembling of ourselves together for thy worship and the preaching of thy gospel. By thy grace we enjoy another of these sacred days. By thy death thou didst purchase for thy people eternal redemption. Thou hast wrought out for them a great and glorious salvation. For thy great love wherewith thou hast loved us thou didst empty thyself of divine glories, and madest thyself a servant among servants, and didst suffer in the garden, and die upon the cross, and enter the grave. Now thou art exalted at the right hand of the Father, a Prince and a Saviour, to give repentance and remission of sins. O thou that judgest men, thy justice is great and glorious as thy mercies. Years ago we tested thy love, years ago we felt the shadow of thy wrath; our guilt made us afraid and we cried unto thee, and thou forgavest our sins, and didst shed abroad thy peace in our hearts. In these recent days thou hast brought other sinners to feel their guilt. They have seen thee upon the cross, and have been smitten with anguish, and have repented, and thou hast received them. Others are bowed down; they mourn; they feel themselves poor and needy; they confess thy justice; they feel the need of thy salvation; they walk in darkness; they grope and find no light; they look unto thee from a distance; but they do not come to thee, they do not follow thee. Wilt thou not draw them to thyself? Wilt thou not bow their pride of heart and turn their wills and make their hearts tender, gentle, and believing? Wilt thou not smite the rock, and cause the waters of penitent grief to flow? Lay thy cross, O Jesus, upon their shoulders and upon their hearts, that they may bear it after thee and share thy glory. Open thou their eyes that they may see eternal destinies and look upon thy divine glories, thy beauty, and thy tenderness. Let them follow thee and trust in thee, strengthened and comforted by thy rod and thy staff. O Christ, for thine eternal love with which thou hast loved us, reach down thine arm mighty to save and lift us up. Lord, save or we perish. And speak thou by thy servant to-day, and cause all that hear to recognize the message not as his, but as thine."

He read as his text Acts xvi. 30: "Sirs, what must I do to be saved?"

He briefly recited the arrest, imprisonment, and release of Paul and Silas. "The salvation for which the jailer cried out was not deliverance from the dangers of the earthquake, nor from the displeasure of the Roman governor. This was the bitter cry of a soul sinking under a load of guilt and trembling at the thought of God's impending wrath. Some of you can appreciate his feelings and his fears. Your sins against God and Christ and the Holy Spirit have risen up before you; they stare you in the face; they condemn you. You feel your guilt—not a light and trifling fault, but guilt deep and dark, such as creatures made in the image of God incur by rebellion against the blessed and holy Creator. The Holy Spirit has recited the divine law in your ears. Your consciences have heard that voice and echoed its condemnation. You desire to escape that divine displeasure; you desire to have the fires of guilt that burn in your consciences quenched. You cry out, 'Men and brethren, what shall we do?' The answer must be drawn from many parts of the Holy Scriptures.

"Understand, in the first place, that you are not to be saved by searching out some plan of salvation for yourselves. Ask for the old paths. 'He that entereth not by the door, but climbeth up some other way, the same is a thief and a robber.' 'Other foundation can no man lay than that is laid.' 'There is but one name given under heaven among men by which we must be saved.'

"Understand also that it is useless to attempt to save yourselves by making yourselves righteous. You have tried, I doubt not, to make yourselves better. Perhaps you have resolved that you would not come to Christ till you can present yourselves in some degree worthy of his care. Have you succeeded in getting rid of your sins? Can you blot out your past sins? Can you erase the record which stands written in the book of remembrance on high? The law of God written in this Bible condemns you; God condemns you; you are condemned already for not believing in the name of God's only begotten Son, the Lord Jesus from heaven. Can you change that condemnation by your feeble, fickle resolutions to reform? 'Can the Ethiopian change his skin or the leopard his spots? then may ye also do good that are accustomed to do evil.'

"Be assured also that it does not belong to you to change your own hearts. 'Ye must be born again;' 'except a man be born again he cannot see the kingdom of God.' But that second birth comes not of blood, nor of the will of the flesh, nor of the will of man, but of God. 'Ye must be born again, but ye must be born of the Spirit.' Notice that the word *saved* is in the passive voice. Sinners do not save themselves; they must be saved by another; they must be saved by one able to save, by one almighty to save, from the wrath of God and from sin, by one able to do for those who trust in him all that they need to have done in order to make their salvation complete and glorious. Christ is able to do this. The crucified and risen Christ is exalted a Prince and Saviour, to give repentance and remission of sins. The word of God says, 'To give,' and he rejoices to give.

"On one point we must pause and dwell with special clearness. Every anxious sinner must not only feel his guilty and lost condition, but he should also thoroughly understand what he means when he asks what he must do to be saved.

He should see to it that he wants that salvation which Jesus gives.

"In the Scriptures the sinner who would be saved is called upon to return to God. He has gone astray. He must retrace his steps. What is meant by this? I mean that man's sin consisted at first and consists to-day in saying, 'I will,' and 'I will not,' in opposition to the will and command of God. God said, 'Thou shalt not;' man said, 'I will.' God says, 'Thou shalt;' sinners say, 'I will not.' If a sinner is to be saved from sin, this opposition must cease. When God says, 'Thou shalt not,' the sinner must reply, 'I will not,' and when God says, 'Thou shalt,' the sinner must answer, 'I will.' The sinner's 'will' and 'will not' must agree with God's 'shall' and 'shall not.' In place of your self-will you must put God's will; that is, repentance, a turning about, a returning to God. But remember, salvation, if it be real and thorough, is not submission for an hour, a day, or a year, but submission for ever and ever. It is submission without condition and without limits.

"The sinner says, 'This is a hard saying,' this utter and boundless denial of self-will and selfishness. But is it hard that the creature should yield to the Creator, that ignorance should yield to wisdom, that selfishness should yield to love, that sin should yield to holiness, that poor, lost, wretched, fallen man should yield to the eternal and ever-blessed God? It is only by yielding that his will is brought into sweet harmony with the will of God, and that he can be a sharer of the divine blessedness.

"Your views on this point should be clear and distinct. If you wish only to be saved from the penalty of your sins, you do not desire the salvation which Jesus gives. He saves his people, not in their sins, but from their sins. If, however, you really wish for his full and glorious salvation, you will desire that your will may be wholly subdued to the will of God. You will be found ready to unite in the memorable prayer of the Lord Jesus, 'Not my will but thine be done.' Salvation implies the giving up of self-will and a reverent submission to the will of God.

"Other sinful passions oppose the grace of God, but chiefly as helpers and supporters of self-will. Pride and vanity strengthen self-will. Turbulent fleshly lusts urge on and back up self-will. Fear of man, fear of danger, and unbelief are but props of self-will. When 'my Lord Will-be-will' submits, the town of Mansoul returns to her rightful allegiance.

"The question at issue between God and the sinner, the question of self-will or submission, is often contested around the performance of some single definite duty. The Holy Spirit often presents to the convicted sinner's conscience some single duty and presses its performance. That duty is a test of the feelings and desires of the sinner's heart. So the Spirit understands it, so the sinner often understands it. As, in the garden of Eden, God gave to Adam a test command, so does he now press upon the conscience of convicted sinners test duties to show them what they are. That which is required may be important, exceedingly important, in and of itself, or it may be in itself of very little consequence, but in every case the duty is all-important and its performance absolutely essential, because the Spirit has laid it upon the sinner's conscience. It will show whether he wishes for salvation from sin or not.

"I used to hear a Christian relate an experience like this. While the Spirit of God was striving with him and conviction of sin was heavy upon him, he felt a clear impression that he ought to go to his barn, and there at one certain place upon the hay-mow kneel and pray. His self-will rose in rebellion, chiefly, it would seem, because it was laid upon his conscience as a duty. But his distress grew upon him. He went to his barn and stood at another place and tried to pray, but no light or peace came; his sense of his sins grew heavier. How could it be otherwise? He went to the spot where he thought that he ought to go, and stood and prayed. Still no peace came, but increasing sense of sin. At length he thought, 'Why should I not? Why not give up my own will? Why not pray that God's will may be done?' He yielded, he kneeled at the place where he had thought he ought to kneel, and there he first felt peace before God. This was a singular experience. Perhaps a man more intelligent and better taught in the Sacred Scriptures would never have such a thing pressed upon his conscience. But the battle of self-will is commonly fought around some single definite duty. That duty may be a confession of wrong done to a neighbor, or conversation with an impenitent associate, or a public confession of sin before the great congregation. Whatever it may be, it shows the sinner his heart and leads him to decide to follow his own will just as he had always been accustomed to do, or it will lead him to pray earnestly that he may be enabled in everything to bow his will to the will of God. He will want the full salvation which Jesus in his grace brings men—salvation from the penalty of sin and deliverance from its power.

"I draw no bow at a venture and speak not doubtfully when I say some of you are standing face to face with duties pressed upon you by the Holy Spirit. Your self-will, supported by pride, and fear of man, and unbelief, and Satanic temptation, refuses to yield. The yoke of Christ seems to you like bondage. The cross is supremely heavy. You draw back from it, and refuse to bear it. I cannot take away the cross which the Spirit bids you bear. I dare not do it; I will not do it. As the messenger of Christ, I repeat the voice of the Spirit and lay the duty, whatsoever it may be, upon your consciences. Do you really and honestly wish to be saved from sin? Then you will yield to the Spirit's kind and gracious movings; you will yield humbly but heartily. If, however, you want something else than the salvation which Jesus gives, what can you expect but perplexity, difficulty, darkness? I beseech of you, deal truly and faithfully with yourselves on this point.

"To those who wish really to be saved I have good news to proclaim. There is a Saviour such as you need. Trust in Jesus as your Saviour. Place the whole work of your salvation in his gracious hands. Christ saves sinners just such as you are. The faith which you are but to exercise is nothing else than your confidence, by which you entrust yourselves to him. Faith has no saving virtue in itself, but it is the hand by which the sinner takes hold of Christ. With this duty few of you will have any great difficulty. When once you wish to be saved from sin and are ready to submit to the will of Christ, you will have no reluctance to take him for your Saviour. You believe that Christ is a divine Saviour. If saved at all, you expect to be saved by him who died on Calvary. Hardly for the world would you resign your opportunity of coming to Christ and receiving his grace. You believe that Jesus is the Christ, the Son of the living God, the great sacrifice for sin. It remains that you should gladly accept what he offers and follow him as loving, trusting disciples.

"Follow the Spirit, and you will be led to Jesus and will come speedily to the joy of salvation; resist the Spirit, and you grope in boundless darkness and fall upon the dark mountains.

"In the Holy Scriptures the question of the text is asked and answered many times. Hardly any two answers are alike. Are there different conditions and different duties required of different men? By no means. But the Holy Spirit adapted the answer to the different spiritual states of the various inquirers. The answer is made to each questioner's heart. A self-righteous young man came to Jesus asking, 'Good Master, what good thing shall I do that I may inherit everlasting life?' Jesus answered, 'Keep the commandments: thou shalt do no murder; thou shalt not commit adultery; thou shalt not steal; thou shalt not bear false witness; honor thy father and thy mother; and thou shalt love thy neighbor as thyself.' The young man answered, 'All these have I kept from my youth up; what lack I yet?' Jesus said, 'If thou wilt be perfect, go sell that thou hast, and give to the poor, and thou shalt have treasure in heaven, and come, follow me.' The young man went away sorrowful. Jesus knew his self-righteousness, and gave him answers which opened that young man's eyes to see himself. He gave him a test command, and the young man's revulsion from that duty showed that, notwithstanding his self-confident claim to righteousness, his riches filled all his heart. If your hearts are filled with the love of the world, you must put your possessions out of your hearts and follow Jesus.

"Nicodemus also came making the same inquiry. He must have asked something like this, for Jesus answered such a question. 'Ye must be born again; ye must be born of the Spirit,' said Jesus. Nicodemus was looking for a legal salvation by outward formal services, but Christ gave him to understand that salvation involves a great spiritual renovation wrought by the Holy Spirit, by which men old in sin become new creatures and enter the kingdom of God as little children. He taught him thus that salvation was only from God. If any of you are looking for a cloak of self-righteous religious duties which you can put on, be assured that true religion springs from a work of God wrought in the heart. You must be born again by the power of the Holy Spirit. You must become new creatures in Christ Jesus.

"On the day of Pentecost the great company of men 'out of every country under the whole heaven,' while listening to Peter's pungent address, cried out, 'Men and brethren, what shall we do?' 'Repent and be baptized, every one of you, in the name of the Lord Jesus, for the remission of sins,' answered Peter. Here were men who had a hand in crucifying Christ, or if they had no active share in that deed of darkness, they had consented to his death; they were partakers of the crime; very likely they had cried, 'Crucify him, crucify him.' They saw their sin, and were pricked in the heart. Well might they repent of their rejection and crucifixion of their promised Saviour, the Son of God, from heaven. Others were devout men who had come to Jerusalem to worship. Like Simeon they may have waited long for the consolation of Israel. How easy for them to enroll themselves among the followers of Christ! All alike are commanded after repentance to put on Christ by baptism. That burial with Christ was the symbol of their dying and living again—of their dying unto sin and living again unto God. The same duties are enjoined upon you. Repent of your long rejection of the grace of God and his Son Jesus Christ, and before God and men devote yourselves to his service by a public confession of Christ in baptism.

"The jailer of Philippi was taken in the midst of his sins. He was holding the servants of Christ in his dungeon. He knew for what offence they had been seized, and he made himself a partner in the crime of persecuting them by the zest with which he thrust them into the inner prison and made their feet fast in the stocks. His conscience was ill at ease. Then came the earthquake's shock, and he felt as if called to stand face to face with his Judge. His soul was pierced through and through with a sense of guilt. 'What must I do to be saved?' he cried in the bitterness of his conviction. 'Believe on the Lord Jesus Christ, and thou shalt be saved,' answered Paul. This is the answer to all of you who are well convicted of sin and have given up all self-righteous hopes. Christ saves you. Look to Christ, ask Christ; whosoever comes to him he will in no wise cast out. Will you not come to him? Will you not trust his promises and commit yourselves to his hands to be saved? He waits to bless you. He delights to be gracious. To save sinners he lived among men, and died and has ascended. His hands are full of gifts. He comes to you, and stands and knocks at the door of your hearts. Will you bolt the door? There is joy in heaven over repenting sinners. This alone of all earthly transactions carries joy to Christ and the angels. Accept of Christ, and earth and heaven will throb with a common joy."

These words were listened to with most earnest attention, for at that time Christ and heaven were realities in the minds of men, and salvation was a living issue. Mr. Wilton spoke as an earnest man, without cant or circumlocution, pressing upon men of thought and conscience the great concerns of eternity. The full result of this discourse will be known only when the opening of the books at the last day shall reveal it, but the beginning of the result was seen in the evening prayer-meeting. When the invitation was given for anxious persons to make known their feelings, both Ansel and Peter arose, and confessing in few words that the Spirit of God had been striving with them, and that they had been resisting the Spirit, said that now they were determined to resist no more, and asked Christians to pray for them that they might be able to submit fully to the Lord Jesus and trust entirely in him. Then there was a pause. Mr. Wilton was just on the point of rising to close the meeting when Mr. Hume rose to his feet. After a sudden start of surprise, a deep hush passed over the congregation, and in the midst of deepest silence Mr. Hume said:

"I have been more than merely an impenitent man: I have been an unbeliever; I have been an infidel. I have not only tried to disbelieve the Holy Scripture, but I have actually disbelieved. I have thought myself wiser than the word of God. I do not mean that I have enjoyed peace, that my conscience has been at rest, and that I have been happy in my unbelief. Three months ago I began to grow more than usually discontented with myself. Questions which I counted settled and put to rest for ever came back to trouble me. A hundred times a day the questions came, What if there be a God who holds me responsible? What if there be a future life and a judgment day? What if Christ be the Son of God? Why such questions should haunt me day and night I could not tell. I have learned to believe that the Spirit of God was speaking to me. This restlessness brought me to the church for half a day. If my object was to gain rest in unbelief, I could not have done worse. My old arguments were unavailing to break the force of the truths preached. The questions which had been sounding in my ears and echoing in my heart began to change to solemn affirmations: 'There is a God;' 'There is a day of judgment;' 'Appointed unto man once to die, and after that the judgment;' 'Christ is risen.' Texts of Scripture learned in my boyhood and forgotten long years ago came back fresh to my memory. But I will not stop to rehearse to you all my struggles of mind for two months past. For a few weeks you have seen me here. I determined that I would try to find Christ if he manifests himself to men in these latter days. For two weeks I have tried to pray, but I have found no satisfaction. Christ has not manifested himself. My darkness has grown deeper and deeper. I have sometimes almost determined to abandon all thought of Christ and throw myself back again upon my former unbelief. But I could not lay down the subject.

"Since I began to try to pray I have felt, faintly at first, like the whisper of a suggestion, but becoming clearer and stronger, like a voice from heaven, that I must in this congregation confess my former state and the feelings which I have had. It seemed to me that I could not do this. It seemed easier to die than to stand up here and confess that my belief, which I had pressed upon others and had boasted of as better than the gospel, had given me no peace. To-day I have been made to understand that the Spirit of God has set me face to face with this confession. I have seen what it means to be saved—that my self-will must die or I must bid adieu to Christ and hope. I cannot live and die hopeless. I cannot rest my head upon unbelief. I confess to you that all my thoughts have been wrong. My beliefs and my unbelief have done me no good. My whole life has been enmity and opposition to the Holy Spirit. I will try to oppose the Spirit no more. I know not what the Spirit may lay upon me, I know not how soon I may break my resolution, but I now feel that I want to be saved from sin, and cannot do otherwise than follow the Spirit though I dwell in darkness for ever. If Christ reject me I cannot complain, but if you think there is hope for one who has so despised the grace of God, I entreat you to pray for me."

It is needless to say that from scores of family altars and closets supplications went up to God that night for the salvation of Mr. Hume and Ansel and Peter, and men prayed especially that Mr. Hume, who for years had been such a tower of strength to the ungodly and the dread of Christians, might be saved for the glory of Christ and the confounding of unbelievers. Those prayers were heard. When the report of that meeting and that confession went out through the community, unbelievers were silent. It was as if the God of battles had emptied his quiver into the hearts of his enemies.

CHAPTER XII.

TRANSFER OF HEAT IN SPACE.

E now turn our attention," said Mr. Wilton, "to a new theme. In the vicissitudes of day and night and of summer and winter heat is transferred *in time*. We now are to look at the arrangements by which heat is transferred *in space*. But since the transfer of heat in space requires more or less of time, the means employed are such as suffice to accomplish both objects. Heat is treasured up and carried away to distant regions, and delivered up for use as occasion demands.

"In a previous lesson the inclination of the earth's axis was spoken of. By this means the northern hemisphere of the earth is turned somewhat toward the sun during one half of the year, and receives a correspondingly larger portion of heat, while during the other half of the year the southern hemisphere is turned toward the sun and is warmed. This inclination of the earth's axis to the plane of its orbit gives us the change of seasons.

"The change of seasons is manifestly designed for the welfare of man. Along with the genial warmth of summer, fruits and grains and the comforts of life are carried far toward the poles, into regions which otherwise would be desolate with perpetual frost. But these extremes need to be softened; otherwise, the violence of the changes would prove destructive rather than beneficent. The severity of these annual changes of temperature is ameliorated by some of the grandest movements and arrangements upon our globe. These arrangements we have in a very imperfect way already examined.

"But there are other inequalities of temperature besides those of day and night, summer and winter. Passing from the equator toward the poles, every degree of the earth's surface passed over causes the sun to sink one degree from the zenith toward the horizon, and gives a corresponding lower temperature, till within the polar circles for a part of the year the sun is entirely hidden and winter reigns without a rival. The temperature of the sea differs from the temperature of the land; the sun comes nearer to one hemisphere than the other, and remains longer north of the equator than south. These and many other differences upon the earth give to different parts of the world every possible variety of temperature and climate. These differences of temperature upon sea and land, from zone to zone and from hemisphere to hemisphere, are equalized or ameliorated by many agencies, but chiefly by a transfer of heat in space, a transfer of heat from place to place.

"I do not need to tell you that while we in the northern hemisphere are enjoying the warmth of summer the southern hemisphere is enduring the severities of winter, and in turn, when winter comes to us, summer smiles upon the nations that live south of the equator. You also remember that the orbit of the earth is not an exact circle, but an ellipse, that is, what is sometimes called in common language a long circle. For this reason the earth is three millions of miles nearer the sun in one part of its orbit than when in another part. Can you tell us, Peter, at what season of year the earth is nearer the sun?"

"In midwinter, or about the first of January.

I have always remembered it because it seemed so strange to me, when I learned it, that the sun should be nearest the earth at the coldest season of the year."

"Yes, one is reminded by it of the humorous argument that the sun must emit cold instead of heat, because when we are at the point of the earth's orbit which is nearest the sun it is winter, and the higher one ascends upon mountains toward the sun, the colder he finds it. But this nearness of the sun while south of the equator would naturally give the southern hemisphere a warmer summer than the northern. For this there is a beautiful compensation. The earth passes through her orbit more rapidly when nearer the sun, and that half of her orbit is also smaller, so that, as the result of this, the sun remains north of the equator about eight days longer than in the southern hemisphere. The sun is nearer while in the southern hemisphere, but the summer is shorter. That which the southern hemisphere gains in distance it loses in time, and that which the northern loses in distance it gains in time.

"The nearness of the sun while south of the equator, the shortness of the summer, and the corresponding distance of the sun and length of the winter would tend to give the southern hemisphere great extremes of heat and cold, a short and hot summer and a long and cold winter. For this also there is a most interesting compensation in the comparative amount of land and water north and south of the equator. Much more than one-half of the dry land lies in the northern hemisphere. This would tend to give the northern hemisphere extremes of heat and cold. South of the equator there is comparatively little land and much water, which tends to give the southern hemisphere evenness of temperature. The inequalities of the earth's orbit and the earth's motion in its orbit we find counterbalanced by the arrangement of land and water upon the earth's surface.

"In connection with this we may notice still another compensation in the elevation of the lands by which the burning heat of the torrid zone and the rigors of the colder zones are more or less diminished. The greater the elevation of any region of country, the cooler must be its climate. Physical geographers like Baron von Humboldt and Guyot have made calculations which show that those grand divisions of the earth which lie in the hot regions of the earth are most elevated above the sea level. South America lies higher than North America, Asia is more elevated than Europe, and Africa is more elevated than Asia. The continents rise as they approach the equator and sink toward the sea level as they come nearer the poles. As these colder lands approach the water level their valleys sink beneath the sea, their coast lines become deeply indented with bays and gulfs, and lakes abound. Thus the warmer waters of the sea are interspersed among the cooler lands, and the temperature of the lands is raised. The very elevation of the continents and the configuration of the lands have a providential relation to the temperature and climate of the world. We cannot suppose that arrangements like these, so aptly fitted to the needs of man, came by chance. In the unmeasured ages past, while this earth was in preparation for man, God had the beneficent *end* in view; nay, in the very beginning, the whole plan and its beautiful completion was had clearly in mind. Millions of ages ago the great Creator tenderly considered the comfort and well-being of the human race, the latest born of his creatures, in these last ages.

"As a general statement, the torrid zone receives an excess of heat, while the frigid zones receive too little, and the temperate zones, lying between, receive, at different times and places, sometimes too little and sometimes too much. The providential arrangements for equalizing temperature are, then, chiefly arrangements for conveying heat from the overheated tropical regions and scattering it over the temperate and polar regions. First among these means we will notice the *trade-winds*, or, as for the sake of brevity they are often called, 'the trades.' Will you tell us, Samuel, how winds are caused?"

"The air is heated at some place and expands; it becomes lighter and rises, while the colder air around rushes in to fill its place."

"You use the words which are commonly employed in explaining the origin of winds, and very likely your idea is right, but the language needs a little correction. The warm air does not rise of its own accord, so to speak, but is pressed upward. The warm air is expanded; it presses outward and upward; the same weight of warm air occupies more space than cold air;

the warm air rises and overtops the surrounding air, and then flows off in order to reach the common level. The column of warm air is lighter than the cooler air, and cannot balance it; consequently, the cold air sinks down, pressing the warm air upward. In this manner an ascending current of warm air is formed, and also currents of cold air flowing from every direction toward the warm centre. These currents continue until the temperature of the air is equalized.

"The atmosphere is commonly believed to be forty-five or fifty miles in height, though some men have estimated its height as very much less than this, while others believe it to be six or seven hundred miles in height. Are we to suppose that the column of heated air reaches to the top of the atmosphere?"

"I think not," answered Mr. Hume. The rarefaction of the lower part of the column renders the whole column lighter than the air around, and the warm air, as we know by the movements of the clouds, after rising a little way, spreads off in every direction, forming upper currents corresponding to the currents below, but moving in the opposite direction."

"Only a few days ago," remarked Peter, "I saw in the same part of the sky clouds moving in exactly opposite directions, and others which seemed to be standing still. I knew how one layer of clouds might be moving north and another layer moving south, but I did not understand why some should be standing still."

"Do you imagine, Peter, that the upper and lower currents of air, moving in opposite directions, come sharply together, the one sliding against the other?"

"I think not," said Peter.

"Supposing, then, as is certainly true, that a stratum of still air lies between the upper and lower winds, does not that explain how certain clouds might be standing still while the others were moving?"

"I might have thought of that myself."

"But how does this carry heat from the warmer region to the colder regions around?" asked Ansel. "I see how the colder air coming in would cool the warm region, and how the warm ascending air would carry away the excess of heat, but how do the cooler regions get the advantage of this heat?"

"That is just what I was on the point of

explaining. Do you remember what was said about the production of cold by expansion and of heat by compression?"

"I remember that if air be rarefied by removing pressure from it, its temperature falls: I think you said that a part of its sensible heat becomes latent; and if air be compressed, its temperature rises. I have seen experiments with the air pump and condenser to prove this."

"That principle explains the transfer of heat by winds. If the heated air rose to the upper regions, and there radiated its heat, nothing would be gained; the heat would be simply radiated into space. But as the warm air rises pressure is more and more removed from it; it expands; its sensible heat becomes latent and is thus kept from radiation; its temperature falls, but not from loss of heat. This rarefied air forms the upper current flowing away from the heated centre. In due time this air must come to the surface of the earth again. Whenever this takes place the air is brought again under pressure; it is compressed, and its latent heat becomes again sensible. Heat is thus transferred from the warmer region to the colder in a latent condition, so that it cannot be lost. We must now apply this to the trade-winds. What are the trade-winds, Mr. Hume?"

"They are regular winds blowing from a little north and south of the tropics of Cancer and Capricorn south-west and north-west toward the equator."

"These winds are called *trade-winds*," continued Mr. Wilton, "on account of their great advantage to trade or commerce. The regular and steady sweep of these winds bears the merchantmen rapidly and safely on their way. The formation of 'the trades' is easily explained. By the intense heat of the sun under the equator the air is greatly expanded and rarefied; the heated air rises along the whole line of the equator; from both sides the cooler air presses in, is heated, and rises; thus steady winds are formed from the tropics, or a little beyond the tropics, toward the equator. If the earth had no rotation upon its axis, these winds would blow directly toward the equator, exactly south and north. The rotation of the earth gives the trade-winds their oblique, south-west and north-west direction. Suppose that a single particle of air at the tropic of Cancer starts upon its journey toward the equator. At its starting it has the same motion eastward as the surface of the earth at that place, that is, about nine hundred and fifty miles per hour. But as it moves on southward the degrees of longitude become longer and the motion of the earth's surface becomes more rapid, till at the equator its motion is one thousand and forty miles per hour. But the particle of air we are watching is not fastened to the earth's surface, and as the earth moves more rapidly the nearer we come to the equator, the particle of air falls behind, that is, the air moves southward and eastward, but the earth moves eastward more rapidly than the air, so that the air falls behind and seems to be moving westward. The result is that the air upon the earth's surface moves south-west. That which takes place with a single particle takes place with the whole body of the air, and that which takes place north of the equator takes place south of it also, producing north-west winds. On reaching the equator the winds from the north and the south meet and stop, forming the equatorial calms, and mingling together, they rise into the higher regions. In rising, the air bears away heat from the torrid zone, and this heat, rendered latent by the expansion of the air, is carried north and south by the upper currents as far as the limits of 'the trades.' In due time these upper currents descend and their latent becomes sensible heat, and is used in raising the temperature. Mr. Hume, can you suggest any method by which we can estimate the amount of heat which is carried north and south by the return trades?"

"I know of no method, except to estimate the amount of heat necessary to raise that flood of air which pours in from the temperate zones to the equatorial heat. That immense amount of heat must, nearly all of it, be carried away to the temperate regions."

"This is the general explanation of the trade-winds. You must understand, however, that, in certain regions and under certain conditions, the trades are liable to interruption or change of direction. Desert regions within or near the tropics give rise to local winds which overpower the trades. In Southern Asia, while the sun is north of the equator, the land becomes so much hotter than the sea under the equator that the trade-wind is overpowered and reversed, forming a wind which blows to the north-east instead of the south-west. But this is only a beautiful flexure, so to speak, of a general arrangement for the greater advantage of a particular region. By this means the summer winds of Southern Asia come from the sea. Northern winds would have been dry. Prevailing northern winds would have made the whole of Southern Asia a desert; but the south-west monsoons come from the Indian Ocean laden with vapor, and render Southern Asia a very garden for fertility.

"The next great agency for equalizing temperature between the torrid and temperate zones is the formation and condensation of vapor. This comes in here, because it depends for its efficiency upon the agency of winds. More than once this method of conveying heat from place to place has been hinted at, but deferred till we came to the proper place to speak of winds.

"The trade-winds, passing over from a colder to a warmer climate, are constantly accumulating vapor. Under the equator the annual evaporation from the surface of the ocean is set down at fifteen feet, or half an inch daily. The formation of this vapor consumes heat which would boil more than eighty feet of ice water. The vapor thus formed is borne upward by the ascending current of heated air. On reaching the higher regions a portion of it is condensed and forms a belt of clouds around the earth. This belt of clouds along the equator is known as the 'cloud-ring.' This cloud-ring shields the belt of calms from the burning rays of the sun and sends down almost incessant rains. But does not that condensation which forms the cloud-ring set free latent heat, and thus intensify the great heat of the equator? Latent heat becomes sensible, but it is given out into the ascending current of air, and serves only to give it another lift till by expansion of the air it again becomes latent. The heat is simply transferred from the vapor to the air. The vapor which remains uncondensed is borne away on the wings of the return 'trades' to the south and to the north, and in due time is condensed and returns to the earth as rain; the heat which is given out by its condensation, wherever and whenever it is condensed, is given over as latent heat to the keeping of the air, and is passed back for use whenever the air descends to the earth.

"Vapor gathered from sea or land is everywhere exerting this equalizing influence upon temperature. Does the temperature rise in any place? Vapor is formed. Every moist body begins to give up its moisture, and the excess of heat is employed in turning this water into vapor. This is the method by which perspiration cools man or beast; whether it be insensible perspiration from the invisible pores of the skin, or perspiration standing in beady drops upon the face of the toiling laborer, vapor is formed and heat is carried away. Have you not noticed on close, muggy days when nothing dries, showing that very little vapor is forming, that perspiration seems to have no cooling effect? It oozes from the skin, but does not evaporate, and hence does not carry off the surplus heat. Animals like dogs and oxen, that do not become wet with perspiration, do not bear heat well; they soon pant and loll, attempting to get rid of the excessive heat through the moist breath and open mouth.

"The sum-total of heat transferred by this agency is too great for comprehension. Look at the Amazon rolling to the ocean a flood broad as an arm of the sea. That great river is brought from the Atlantic Ocean on the shoulders of the trade-wind. As the vapor is slowly lifted by the rise of the land from the sea level to the summits of the Andes, it is condensed, and falls as rain. Well is it for South America that the Andes were thrown up on the western coast, for the winds west of the mountains are dry as a pressed sponge, and the most of that narrow slope is barren and desolate. South America would be a desert if the Andes ran along the eastern coast. Look at the Mississippi, and the great rivers of Europe, and the matchless rivers of Southern Asia. All the rivers of the world represent only the *wastage* of the rain which falls upon the land after supplying the wants of the vegetable kingdom and keeping the lands moist. All this water is lifted into the air by heat, and every movement of vapor is a movement of heat. Every particle of vapor goes freighted with heat. Every cloud driven across the sky represents the transfer of heat, and every transfer is in the direction of equalization. Everywhere the tendency is to equilibrium. Nature has no processes for transferring heat from colder to warmer regions.

"We may form a conception of the amount of heat transferred by the agency of vapor by estimating the amount of heat-force required to evaporate the water which forms our rain-clouds and lift them into the upper regions. According to a calculation of Mr. Allen, late of Providence, to evaporate one-eighth of an inch of water daily from that belt of the surface of the earth lying within the tropics, and raise it five thousand feet high, requires 4,700,000,000 horse-power, or one hundred and thirty times the effective force of the whole human race, reckoning it at 250,000,000 able-bodied men. But the actual evaporation from the sea within the tropics is believed to be about half an inch daily— four times as great as Mr. Allen's supposition.

"I see, however, that our time is nearly exhausted, and I wish before closing to revert to that more important theme upon which I spoke this forenoon. I do not know how the truths preached interested or affected you, nor do I now wish to have you tell me. I wish only to say that, as the sermon was preached at your request, I hope it proved applicable to you, and that you will give the truths presented earnest attention. Consider them well, and make your conclusions known this evening."

The conclusion which the evening made known, you, reader, have already learned.

CHAPTER XIII.

OCEAN CURRENTS AND ICEBERGS.

 WEEK has passed since Mr. Hume made his frank confession. He went home no lighter of heart than before, yet he felt in some respects different, for he had attempted to do what was right in the sight of God. But he did not feel the joy of sins forgiven. He had not looked upon Christ as a Saviour for himself. He felt that God had distinctly set life and death before him. His doubts were gone; the spiritual world was a reality; Christ stood at his right hand and Satan at his left; he stood where the path of destiny divided, the one path leading up to heavenly seats with Christ, the other leading down to darkness and despair. A voice seemed to be whispering in his ears, "This is the last call." He went to his chamber determined, if possible,

to settle the question of life or death before he left the place and before he slept. He took his Bible, and on his knees turned and read the Psalms at random. But the cloud of darkness only gathered deeper. The words of David's penitential Psalm caught his eye: "Against thee, thee only, have I sinned, and done this evil in thy sight." He felt that these words of David were true in his case also. All his long impenitence and bold unbelief had been against God. By night and by day, for many a long year, before the sleepless eye of God, he had lifted up his hand, almost defying the holy One, yet the lightning of God had not smitten him. He wondered as much at the long-suffering of God as at his own dreadful daring of the divine wrath. He had been taught better things; he was trained to know the Scriptures and to go reverently to the house of God, but he had turned from Christ and hope. He read on: "Deliver me from blood-guiltiness, O God, thou God of my salvation." He felt that this belonged to himself more than to David. David had shed the blood of natural life, but he had destroyed the souls of men. He had stood chief among unbelievers. He had led young men into infidelity. He had seen them drink in his unbelief like water, throw off all restraint, and rush headlong to ruin. He had wrought a work of evil which he could never undo, and for which he could make no atonement. What was a confession in comparison with the ruin he had caused? What could his confession do for the young men already, perhaps, among the lost through his influence? Could his late repentance call them back to life and hope? Would God forgive and raise to heavenly heights a man who had dragged others down to hell? Would it be possible that Christ should fill his soul with blessedness while his victims were drinking the wine of the wrath of God? A deep horror seized him. The darkness of eternal death seemed to enfold him. Must he, then, after having caught a glimpse of life and joy, be cut off from hope and be driven from God for ever? This would be just, but he felt that he could not endure it. "O thou great and holy God," he prayed, "I will ascribe righteousness to thee though thy righteous wrath shall sink me to hell; but, O thou merciful God, my soul cannot endure thy justice. The foretaste of thy wrath fills me with the pangs of eternal death. O God, have mercy upon me. O God, blot out my transgressions. Create in me a clean heart, and renew a right spirit within me. O Christ, whom I have despised, cast me not from thy presence. Help me to submit to thee. Help me to follow thee. Spare me that I may undo something of that which I have done against thy glory and the souls of men. O Jesus, I can do nothing to save myself. O Lord, have mercy on me, the chief of sinners."

He read the invitations and promises of Christ, and prayed again. Again he read and again he prayed. Little by little the promises of Christ stirred a feeble faith in his heart; he felt that there was still hope for him, and with the determination to cast himself upon the sure mercies of Christ and to devote himself to his service, he threw himself upon his bed, and being wearied almost to exhaustion, soon fell asleep. When he awoke it was broad daylight. He had slept a sweet, refreshing sleep. But he was refreshed not merely in body. He woke to a new world. His heart was filled with sweet thankfulness. "How beautiful," he said, "is God's world! I never saw it so before, but the earth and sky seem clothed in glory. But most wonderful of all is God's goodness to me. I have rebelled against him all my life, yet he has loved me and sought for my salvation, and now the sunlight of his love has broken through the thick clouds of my sin, and a day of hope and joy has dawned upon my life. Christ has indeed revealed himself. Blessed be his holy name for ever and ever! What shall I render unto the Lord for all his benefits? I will take the cup of salvation and call upon the name of the Lord. I will pay my vows now in presence of all his people. I will teach transgressors thy ways, and sinners shall be converted unto thee."

All this was known to the people, for during the week Mr. Hume had spoken of it in private and in public. He had told it to Mr. Wilton, and they had rejoiced together.

Ansel and Peter had also regularly presented themselves at every meeting as anxious inquirers desiring the grace of God. Peter had also on his knees said from the heart, "Here, Lord, I give myself away," and had received the assurance that his sins were forgiven. The Spirit of God witnessed with his spirit that he was born of God. He began at once to use all his influence to bring his young friends to Jesus. The addition of two such workers as Mr. Hume and Peter, each moving in his own circle of acquaintances, gave a fresh impulse to the religious interest, which was now becoming deep and pervasive. Especially had Mr. Hume's conversion, so clear and positive, confounded those who had sat "in the seat of the scornful," and many came in now for the first time to see for themselves what it could be that had mastered their cold, clear-headed leader in unbelief.

But Ansel still walked in darkness. He had talked with Mr. Wilton, but no light had entered his mind. He said that he thought he had submitted in all things to the will of God. He was becoming impatient that Christ had not come to him as to others. This was their condition as they came together upon the Lord's Day. They all understood each other, and had no need now to ask questions or make explanations. Mr. Wilton believed that the study of God's works would not interrupt the working of the Holy Spirit, and therefore went on with his lesson as usual.

"We have already spoken of the transfer of

heat from the torrid to the temperate and frigid zones by the agency of winds and watery vapor. These carry heat chiefly in a latent condition. But great movements of heat take place in a sensible state. In this transfer of heat, also, water is the great carrier. The winds and vapor go freighted with latent heat above, and the waters and wind go freighted with sensible heat below. We will first examine the operation of the ocean currents.

"Not only do rivers run through the lands and hasten to the sea, but in the midst of the oceans rivers are flowing in comparison with which the Mississippi, the Amazon, and the Yang-tse-kiang are rippling brooklets. The earth is belted by these ocean streams traversing the seas." An ocean current, called the Gulf Stream, issues from the Gulf of Mexico between the Florida coast and the Bahama islands. It flows northward off the coast of the United States, gradually increasing in breadth and spreading over the Atlantic Ocean. It is deflected by the New England coast and the great shoals off Newfoundland, called the Grand Banks, or else by another current flowing southward from Baffin's Bay, and strikes across the

North Atlantic, bathing the shores of the British islands and reaching even to Iceland.

"The general outline of the ocean currents is this: issuing from the South Pacific, a current flowing eastward splits upon Cape Horn. The western portion, called Humboldt's current, flows northward along the western coast of South America, and is swallowed up and lost in the great equatorial current of the Pacific. This is a broad current flowing westward and covering the entire space between the tropics. Striking upon the eastern shores of Asia, this equatorial current divides, one part flowing northward along the coast of Asia, the other finding its way through the many islands, sweeping across the Indian Ocean, and flowing down the eastern shore of Africa on each side of Madagascar. Doubling the Cape of Good Hope, the current continues in a north-westerly direction across the Atlantic. Striking upon Cape St. Roque, this current again divides; a part flows south and a part pours into the Caribbean Sea. From the Caribbean Sea it issues as the Gulf Stream, of which I have already spoken. This Gulf Stream impinges upon the western coast of Europe, and pours partly into the North Sea and partly flows south off the western coast of Africa, completing thus the circuit of the Atlantic. The currents of the Indian and of the great Southern Oceans are as yet very imperfectly understood. Of all the ocean streams the Gulf Stream is most famous and best understood. I shall therefore use this as an illustration of the agency of ocean currents in conveying heat and modifying climate.

"The waters of the Caribbean Sea are heated by the tropic sun to eighty-eight degrees. From these heated waters the Gulf Stream issues salter and warmer, and of a deeper blue, than the waters of the surrounding sea. Its greatest velocity as it issues from the gulf is a little more than three miles per hour. As it flows northward its velocity diminishes, its breadth becomes greater, and its depth less. It covers thus with its warm waters a broad belt of the Atlantic Ocean, and extends its influence to the most northern part of Europe. You can judge of the amount of heat which is removed from the tropics when I tell you that the unmeasured flood of the Gulf Stream would swallow up three thousand rivers like the

Mississippi. This one ocean stream is many times greater than all the rivers of the world. We feel the warmth of the Gulf Stream with every wind that blows from the sea. To this the British isles owe their mild, moist climate and perennial greenness, and by its influence a winter in Iceland, upon the Arctic circle, is no more rigorous than a winter in Montreal, twenty-one degrees nearer the equator. But what is the Gulf Stream, though it be fifty fold greater than all the rivers of the world, in comparison with the whole sum of the ocean streams? Upper currents and under currents fill the sea. They meet the explorers of the sea everywhere. The navigator drops his measuring line, and finds it swept away and drawn out by unseen currents. All these movements of the waters are in favor of the equalization of temperature. The cooler waters of the frigid and temperate zones are mingled with the heated waters of the tropics and exchanged for the equatorial waters. The transfer of heat would not be greater if broad rivers of molten lava were flowing from the equator to the poles.

"Another agency for the transfer of heat is

the movement of ice, and especially of icebergs."

"Will you not tell us," said Samuel, "how these ocean currents are produced? I can understand how winds are formed, but I do not see that these streams in the sea could be formed in the same way."

"I designed to speak of this, but for the moment it had slipped from my mind: I am glad that you called my attention to it. I do not expect, however, to give a full and satisfactory account of their origin. If I should do this, I should succeed where every other man has failed. I shall not attempt a full explanation. By some means or other, the waters of the ocean are thrown out of equilibrium, and these currents are plainly an effort to restore the balance or equilibrium of the waters. Many influences and agencies conspire to disturb the equilibrium of the sea. The attractions of the sun and moon are constantly counteracting the attraction of the earth and lifting the waters, so to speak, above their natural level. The tides produced by these attractions of the sun and moon are the immediate cause of some of the minor local currents. The winds set the waters in motion, tending to pile them up in one place and leave the sea below its natural level at another. The effect of strong winds in piling up the waters, even upon our great lakes, is very considerable. A heavy east wind upon Lake Erie has been known to drive the waters toward the western end of the lake so much as to leave Niagara River above the falls almost dry. On the other hand, a heavy west wind drives the waters eastward, and produces almost a flood in the river. The influence of constant winds like the 'trades' acting upon an immense expanse of water must be very much greater. Unequal evaporation tends to destroy the balance of the waters. In the colder regions the evaporation is very little, while within the tropics it amounts to about half an inch daily, or fifteen feet per annum. The head of the Red Sea is two feet lower than its mouth on account of evaporation. This unequal evaporation causes also an unequal saltness, and consequently an unequal weight. The fresher and lighter water cannot balance an equal bulk of salter and heavier water. When once currents are started the revolution of the earth upon its axis would affect them, just as the rotation of the earth affects the trade-winds. Now, all these various agencies, and perhaps many others, combine their influence to destroy the equilibrium of the waters of the ocean. They unite and interweave their influence in a thousand ways beyond all human calculation. The result is the ocean currents. But how much is due to one cause and how much to another in the present state of knowledge no man can tell. Only for a few years have the phenomena of ocean currents been made the object of scientific observation and research. But the effect of ocean currents in modifying climate is well understood, and the modification of climate means nothing else than the transfer of heat. This is all that I have to say of the rivers of the sea, and if there are no more questions, we will now look at the movement of heat caused by icebergs."

No question was asked, and Mr. Wilton continued:

"In polar regions there must be an immense formation of ice. Except in the oceans, the movements of water are chiefly movements of water in the condition of ice. Only for a small part of the year could water exist unfrozen. Immense regions of the Antarctic continent seem to be covered with one broad glacier. The ice pushes down into the sea until, undermined by the dashing of the waves, it breaks off, and enormous fragments are launched upon the deep waters. Sir James Ross saw in the southern ocean a chain of such icebergs extending as far as the eye could reach from the mast-head, many of them from one hundred feet to one hundred and eighty feet in height and miles across. Captain d'Urville saw one thirteen miles long and one hundred feet high. Its bulk was so vast that though the waves were dashing against it not a tremor was perceptible. Astronomic observations could be made from it as if it were solid rock rooted in the heart of the earth. In the same manner icebergs are formed in the northern ocean also. How much heat is given out in the freezing of water?"

"About one hundred and forty degrees," answered Peter.

"In the formation of icebergs, then, heat is given out nearly sufficient to boil an equal quantity of cold water. The icebergs float away toward the equator. They come down from Baffin's Bay till they meet the Gulf Stream off Newfoundland. In the southern hemisphere they come ten degrees nearer the equator. As they float toward the tropics they slowly melt, and in their melting they exact from the air and the sea where they melt the same amount of heat which they gave up in their freezing. If they melted at the same place where they froze, there would be no transfer of heat. But they are formed in the polar regions; they give out their heat in the frigid zone, while they melt and absorb a like amount of heat from the temperate zones. In this manner the polar regions are exchanging with the temperate zones ice for water. They borrow water, rob it of its latent heat, and send it back in the form of ice. The temperate zones supply the needed heat and bring the ice back to the form of water, when the polar regions again borrow it, seize upon its heat, and again send it back in the form of ice mountains. The effect is the same as if thousands of railroad trains were transporting water to the frigid zones, leaving it there to freeze and give up its one hundred and forty degrees of latent heat, and bringing it back in the form of ice. Let us estimate the bulk of one such iceberg as that seen by Captain d'Urville. It was thirteen miles long and one hundred feet high, and we will suppose that it was four miles broad. Standing out from the water one hundred feet, it must have sunk at least eight hundred feet below the surface. This would give us the enormous bulk of (1,304,709,120,000) one trillion three hundred and four billions seven hundred and nine millions one hundred and twenty thousand cubic feet of ice. The burning of one pound of coal will generate heat sufficient to melt about five and a half cubic feet of ice. To melt one such iceberg would require more than one hundred and eighteen millions of tons of anthracite coal. This is the amount of heat given out in the polar region by its freezing. This is the amount of heat transported from the warmer to the colder regions. But what is one iceberg to the thousands which drift yearly from the frigid zones toward the tropics?

"But even this hardly represents the entire transfer of heat by the agency of icebergs. The icebergs are formed from the snows of polar storms, and these are formed from the condensation and freezing of vapors. In the process of condensation one thousand degrees of heat are given out. Every iceberg *represents* a transfer of heat sufficient to boil more than six times its weight of ice water.

"One marked illustration of the effect of icebergs we ought to notice. Down through Baffin's Bay icebergs are constantly floating. They are borne on southward till, in the still waters of the Grand Banks, between the polar current and the Gulf Stream, they float around and melt and disappear. To these melting icebergs the chilliness and unfailing fogs of the Grand Banks are due; and not only this, but the very existence of the Banks is supposed to be due to the deposit of sediment, sand, earth, and stone brought by polar ice.

"I have spoken only of the polar glaciers and the icebergs formed by their pushing off into the sea. But the same transfer of heat is taking place, on a very much smaller scale and within narrow limits, by the glaciers of the Alps and every other mountain glacier. The glaciers are nothing else than rivers of ice. Snow falls upon the mountain tops and valleys of the mountain sides from age to age. The snow slowly changes to the structure of ice, and by its enormous weight flows down through the gorges of the mountain sides, till in the warmer

Curiosities of Heat.
Page 288.

vales below it melts and disappears. We have not time to go into a full examination of all the interesting phenomena of glaciers, but this one point you will notice and remember: these rivers of ice—for they flow like rivers—cool the valleys and tend to warm the mountain tops; of course upon the tops of the mountains there can be no accumulation of heat, because, standing out into the eternal coldness of space, and swept by winds for ever, and exposed by the thinness of the air to a rapidity of evaporation unknown at the sea level, heat is caught up and borne away in a moment.

"This closes this department of our theme. I might have gone much more into details and given you great stores of particular facts and figures, but they would have added nothing to your understanding of the subject, and we can hardly afford to devote our Lord's Day to mastering the details of the natural sciences. We have now looked at some of the methods by which the extremes of heat and cold, in day and night, in summer and winter, and in the tropics and polar regions, are mitigated. The same principles operate upon the smallest and upon the largest scale. If there is need for me to attempt in a formal way to awaken in you admiration for the wisdom and goodness of God shown in all these beneficent arrangements for equalizing temperature, our study has been largely in vain. We have only to remember that all these contrivances are the Lord's designs. He created the world; he endowed matter with its qualities and forces, and he gave it these qualities and forces for the purpose of using it as he has used it. He planned all those contrivances by which he secures the comfort and the good of man, and the fact that these natural agencies are fitted for moral uses in recovering sinners to holiness and blessedness is but the culmination of its adaptation to the uses of man.

"This, however, does not complete our course of study. A few other points will demand our attention for two or three more lessons. But while we go on with our studies of Nature, remember that the physical was created for the sake of the spiritual; the spiritual is more important. Let us not subvert the divine order and sink the high purpose of the creation to mere material agencies and contrivances. To know God is greater and better than to understand Nature. That we might know and enjoy and glorify the Creator was the object of our creation. We cannot express it in better language than that employed in the old catechism: 'The chief end of man is to glorify God and enjoy him for ever.' That term 'for ever' includes the present life as well as the future. We ought to know, enjoy, and glorify God to-day. I hope that another week may find Ansel with some happy experience in this matter."

CHAPTER XIV.

COMBUSTION.—COAL-BEDS.

ANOTHER Lord's Day comes, and no change has taken place with the class which calls for mention. Ansel still walks in darkness, ready indeed on every occasion to manifest his concern for the salvation of his soul, diligent in reading the Scriptures, frequent in prayer, and giving yet no indication of a flagging of his avowed purpose to follow Christ, but he receives no comfort and peace. A painful and distressed interest is becoming more and more concentrated upon him. What will be the end of his groping in darkness? This cannot last always. Unless the hindrance, whatever it be, which prevents the exercise of faith, be seen and removed, Ansel will probably soon go back to his former careless state, and, it may be, become tenfold more obdurate than before. He will be likely, on the one hand, to become self-righteous from his supposed effort to come to Jesus, and, on the other, discouraged and despairing, feeling that for him effort is vain and salvation unattainable. While he remains in this state the very lapse of time is dangerous. All feel concerned for him, but no questions are asked, and the lesson goes on as usual.

"The method of transferring heat which we are now to examine is wholly different in principle from any which we have as yet considered. I refer to the production of heat by combustion. The transfer of heat by combustion cannot be compared for vastness with those great movements of heat which have before claimed our attention, yet for the comfort and well-being of the human race combustion is exceedingly important. Without that command of heat which combustion gives, man could not rise at best above the savage state, and in fact could hardly exist upon the earth. We smile at the Grecian myth that Prometheus stole fire from the gods and brought it to men in his reed staff, but fire is certainly worthy of being counted one of God's great gifts. But whence comes the heat of combustion? Is it a new and original generation of heat, or is it merely a transfer? Will some one explain this?"

"I don't think that I can tell," said Samuel. "I remember the principles you have given us about the nature and production of heat, but I do not know how to apply them to combustion."

"I did not suppose that you would be able to explain all the phenomena of Nature at sight, yet the production of heat by combustion is not difficult to be understood. The burning of wood and coal is chiefly the union of oxygen with carbon. The oxygen of the air unites with the carbon of the combustible. The attractive force between oxygen and carbon is very strong. When they unite, the atoms of oxygen dash against the atoms of carbon with great violence. As they dash one upon another their motion is lost, but by the laws of transmutation of forces that lost motion reappears as heat; that is, the motion of the atoms as they fall the one against the other is changed to that vibration of the atoms which we call heat. The atoms of carbon, in their separation from oxygen, may be compared to weights suspended, ready to fall. Let once the cord be cut, and the weight falls and dashes against the earth; its motion in falling is lost, and reappears as heat. So carbon is suspended, so to speak, waiting to unite with oxygen. But how is the weight raised? How is carbon brought into this state of suspense, waiting to dash upon oxygen and develop heat? That is not its natural state.

"Carbonic acid is found everywhere mingled in small proportions with the atmosphere. This carbonic acid is nothing else than carbon and oxygen united in the proportion of one atom of carbon to two atoms of oxygen. This is the natural state of carbon. This carbonic acid is the food of plants; it is this which supports all vegetable growth. The carbonic acid is absorbed by the leaves of plants and trees, and in the hidden laboratory of the leaf, by what process is one of the undiscovered secrets of Nature, the carbon is separated from the oxygen, the oxygen is discharged through the pores of the leaf, and the carbon is carried into the circulation to build up the fabric of the woody fibre. That which the most skillful chemist in the world cannot do, except by indirect processes and at a high temperature, the leaves are doing directly at the ordinary temperature. Vegetable growth is a deoxidizing process. To accomplish this an enormous force is requisite. To separate carbon and oxygen, a force is demanded which is able to overcome their powerful attraction. How shall we estimate the strength of this force? In order that they may unite, as in the explosion of gunpowder, solid rocks are torn asunder. The attraction of carbon and oxygen is strong enough to tear great rocks in twain. It is this attraction which sends the cannon ball and the shell like meteors of death upon their errands of destruction. This great force must be overcome; carbon must be separated from oxygen and built into trees. This is the lifting up of the weight. But whence comes the force necessary to accomplish this? From the sunbeam. The heat of the summer's sun, employed as force, is used to deoxidize carbonic acid. Heat is used, and used up, in lifting the weight which in its fall shall generate again a like amount of heat. The combustion of wood produces the same amount of heat as was needful to separate its carbon from the carbonic acid of the air. Vegetable growth is thus a cooling process; heat is withdrawn from use as heat, and is employed as force. As force it has nothing to do with temperature. The summer's heat, employed in vegetable growth, reappears in the blazing billets of the kitchen fire. Heat is condensed and solidified, so to speak, and placed under man's control. In this solidified form heat may be laid up in store or transported at pleasure.

"The grandest application of this principle is seen in the formation of the coal-beds. At some early period in the unmeasured ages past, the temperature of the earth must have been much higher than it now is; the air was filled with moisture, and carbonic acid abounded. As a consequence, there was an enormous vegetable growth. This, as we have seen, is a heat-consuming process. The heat is withdrawn from the air and employed in deoxidizing the carbonic acid. This vast vegetable growth—enormous ferns and coniferous trees—fell, and was swept by rivers or by floods into valleys, or the beds of lakes, or the sea; the sediment of the waters covered it, and there, shut up from the air and subjected to a heavy pressure, this vegetable mass underwent a slow transformation. Peter, have you ever seen a coal-pit? I do not mean a coal *mine*, but that which charcoal-burners call a coal-pit."

"I have seen them many a time."

"Tell us, then, how wood is burned to coal without being burned up."

"The wood is set on end, closely packed in the shape of a mound, and then covered with earth. Fire is kindled in the middle of the pile, and just enough air admitted through air-holes at the bottom to keep up a slow burning. It burns just fast enough to heat and dry the wood without burning it up."

"The same process," said Mr. Wilton, "went on in the formation of the coal-beds, but very much more slowly. Under the pressure of earth and water the vegetable deposits lie smouldering, not for a few days, but probably for ages, till nothing but the carbon remains, and that pressed into a solid mass heavy as stone. Veins of coal are found interspersed with layers of earth and rock, layer above layer, and these layers are commonly not level, but more or less inclined and sometimes broken. This shows that a deposit of driftwood was made, then a deposit of sand or clay, then another deposit of vegetable material and another layer of earth. At length, by internal convulsions, the whole surface was raised from beneath the waters, and in due time the coal-veins were laid open, and the coal brought out for the use of man. Then the force so long pent up and held in suspense is set free; the stored-up heat of the geologic ages is brought out for use. The excess of heat in that ancient period is handed down to these later times. How sublime this transfer of heat! It carries us back, in imagination, to the 'heroic ages,' so to speak, of the history of creation. By other methods heat is treasured up for a day or a year: by this method it is kept in store for myriads of ages. We see that the same natural forces were working in those early ages as to-day, and the same benevolent Creator was arranging the affairs of the world for man's advantage. The sunbeam which streamed upon the earth long ages before man was created is to-day smelting ores, driving machinery, dragging ponderous trains of loaded cars, and ploughing the seas with freighted keels. This seems like a fairy-story or a dream, but instead of that it is the soberest of philosophic and scientific truth.

"We ought also to notice the internal heat of the earth. This has been handed down from the day of creation, it would seem, till the present. No new principle is seen in the earth's internal fires, but a sublime illustration of the storing up of heat in a hot body and its slow radiation.

"The origin of the internal heat of the earth we can only conjecture. Perhaps God created the various elements separate, uncombined, and allowed them then to combine according to their natural affinities. This sublime conflagration of all the elements of the earth would generate the highest temperature which could be produced by combustion. The elements would melt with fervent heat; everything which could be vaporized by heat would be turned to vapor. Then radiation of heat would begin. Vapors would sink to fluids and fluids turn to solids; a hard crust would be formed on the surface of the globe through which the heat of the still molten mass within would be slowly conducted and escape. Upon this internal heat the earth depends in no small degree for its temperature. The heat generated perhaps upon the day of creation helps now to render the earth habitable.

"That the earth was once in a fluid state and has lost a portion of its heat by radiation is indicated by several facts. It is one of the received beliefs among geologists that at some period in the past the temperature of the earth was much higher than it now is. The animals and plants which flourished during the ages when the coal-fields were deposited show that sea and land were warmer than at present. It is believed that the change of temperature has taken place on account of the cooling of the earth from radiation. The rate of radiation is so slow, however, that no farther sensible change of temperature can take place for thousands of generations.

"The form of the earth also indicates that it was once fluid. The earth is an oblate spheroid, a flattened sphere, and has that degree of flatness which a fluid mass would assume if revolving at its present rate. The earth swells at the equator and rises thirteen or fourteen miles above the sea level at the poles. The waters of the ocean move freely and take the same form as if the whole globe were fluid, and the solid parts of the earth have the same degree of convexity, which shows that it took its form from its own rotation upon its axis while in a fluid state. This would also show that in the primal ages, when the earth was in a plastic or fluid state, it had the same rate of rotation as at present.

"The lifting up of the mountain ranges also

is best explained by supposing that the earth was once molten. The earth cooled, a crust was formed, and by farther cooling and contraction of the molten mass within the crust wrinkled and formed mountain chains. Thus the higher temperature of the geologic ages, the form of the earth as if it were a revolving fluid mass, and the corrugation of its surface—these, joined with its present internal heat, point to the fact that it was once molten and fluid to its surface. The benefits of this heat laid up in store on the day of creation we still enjoy."

"Before the class is dismissed," said Mr. Hume, "I should like to say a few words."

"I have nothing farther to say to-day," answered Mr. Wilton, "and we should be glad to hear you now. Say on."

"I wish only to say that these lessons have led me to such thoughts of God's wisdom and goodness as I never had before. Of course it is not strange that this should be the case with me. I now look at everything with new eyes. It is not merely this one element of heat in Nature that moves my admiration, but I have been led to consider a thousand things in which the goodness of God is shown. My thoughts of the divine goodness are as fresh and interesting to me as my impressions of his righteousness and holiness are startling. For years I have tried with might and main to look upon the dark side of the world and to exaggerate its physical evils. I have searched for disorder and want of adaptation. As long as I misunderstood the purpose of the creation, I thought I was successful in impugning the wisdom of the arrangements of this physical world. While I supposed that the earth must needs be the Creator's masterpiece in beauty and pleasantness and all manner of perfections, designed just to give sensual pleasure to its inhabitants, I could find, or thought I found, many faults in the Creator's work. Now I withdraw all my former charges. My eyes are opened. The rougher elements of man's life will henceforth have a new meaning to me. I see that God seeks not so much present pleasure for men as their holiness. He lays a solid foundation for their happiness. He seeks to render men blessed by bringing them into likeness and union with himself. These are new views to me, and I thank my heavenly Father that this new light has dawned upon me. I feel now that I can bear the ills of this life cheerfully, understanding that the Lord is using them as a means of spiritual discipline. It seems to me as if this lower world and man's lowly life were already glorified by a beam of light falling from heaven. I hope that my young friends have been as much profited as I have been."

"I rejoice with you, Mr. Hume. 'We know that all things work together for good to them that love God.' This light has shone upon me for many years."

CHAPTER XV.

ECONOMY OF HEAT.

"IN this final lesson I wish," said Mr. Wilton, "to bring before you some general views of the whole subject of the agency and management of heat.

"When Jesus had fed the five thousand men upon the mountain side by the Sea of Galilee, he said to his disciples, 'Gather up the fragments that remain, that nothing be lost.' The Christ who spoke these words was the same Christ by whom 'all things were created that are in heaven and that are in the earth, visible and invisible.' These words inculcate the propriety of saving, the very opposite of extravagance and wastefulness. The same prudent economy we find in all God's works. Nothing is wasted. God provides bountifully; he is not stinted in his works; we find nothing narrow or mean; his resources are ample for all his undertakings. Perhaps a careless observer might charge him with prodigality and wastefulness. The wilderness rejoices in beauty and fertility upon which no human eye gazes, and which supplies no human want.

> 'Full many a gem of purest ray serene
>   The dark unfathomed caves of ocean bear;
> Full many a flower is born to blush unseen,
>   And waste its sweetness on the desert air.'

Rich fruit grows ruddy and golden in the autumnal sun only to fall and decay. How small a part of the seeds which might germinate and reproduce the parent plant ever fulfill this their legitimate object! But this is not waste. As for the beauty with which the unpeopled wastes are smiling, we know not what other beings besides man 'grow glad at the sight.' Fruits and grains and seeds were appointed as much to nourish the animal kingdom as to reproduce plants and trees. And that which decays is not wasted. The oak lifts high its leafy arms and does battle with the tempests for a century, and then having served its purpose in Nature, if man does not call it to the higher mission of serving his purposes, Nature begins to pull down the structure she has reared and rebuild the elements in other forms—such forms as man perchance may need. The fruit that falls and decays is not wasted; it shall blush with golden tints in other forms and in other years. God pulls down the old that he may build the new. The same elements appear and reappear in a thousand shapes. There is endless change, but no waste. This sentiment, 'Gather up the fragments, that nothing be lost,' which is proclaimed throughout all Nature, is uttered most emphatically in the management of heat. God has provided most bountiful stores of heat, but has left no heat to go to waste. Will you, Mr. Hume, suggest one of the general arrangements for the economical use of heat?"

"I think that the arrangement for economizing heat which ought to be mentioned first is the confinement of heat to the locality where it is needed."

"Will you explain that a little farther, Mr. Hume?"

"All living creatures are confined near the surface of the earth. They penetrate only a few feet into the earth and soar a few hundred feet above it. Heat is therefore confined to the region of the earth's surface. It penetrates but a little way below the surface, and when warm air rises into the higher regions, heat becomes latent. The higher parts of the atmosphere are cold, and in the empty spaces of the heavens the temperature is we know not how low. God has provided for heating only that part of the world which needs to be heated. I think you spoke of this in some one of the earlier lessons."

"Perhaps I did. But I refer to it again to call especial attention to the idea of the economical use of heat. Who will mention another method by which heat is economized?"

No one answered.

"I asked the question, but did not expect an answer. God shows economy in the use of heat by accomplishing many different results by its agency. I do not mean that the same identical heat accomplishes different results at the same time. The same force cannot accomplish two works. As man cannot spend his money and at the same time keep it, no more can heat be used and not used up in that form. The heat which raises the temperature can do nothing else at the same time, and when it is employed as force it ceases to affect temperature. But by this one agency of heat the Creator brings very various works to pass. Heat expands bodies, relaxes cohesive attraction, and brings the chemical affinities into activity. By this means the elements of Nature are subdued to human uses, seeds germinate, all the processes of vegetable life go on, and digestion and nutrition are carried forward in the bodies of animals. By the agency of heat the winds blow, the deep waters of the ocean circulate, clouds are formed, dew and rain refresh the earth, rivers flow, and all the activities of life fill the world. The employment of one agency for the accomplishment of so many works indicates economy in the expenditure of force and means. Moreover, the same heat appears and reappears again and again, passing from the sensible to the latent form and back again, asserting itself alternately in raising the temperature and as active force. A beam of heat falls upon our world: it is partly absorbed by the earth, and warms it. A part of that warmth is used in setting the chemical affinities in action in the sprouting of seeds; a part warms the air by conduction; a part is radiated, and being stopped by the vapor in the air, warms it; the heat of the air is partly used in the evaporation of water: the vapor formed is condensed and waters the earth, and gives out the heat by which it was formed; that raises the temperature of the air; a part of it is used in deoxidizing carbonic acid and building up the forests; the forest tree falls by the woodman's axe, is burned for fuel, and gives out its heat again, or if it falls and decays, the result is the same; the heat given out by combustion cooks the laborer's dinner and warms his room, or it goes out again, and is used in preparing food for the growing wheat; that wheat is used for food, and by slow combustion in the blood the heat is again evolved, the body is warmed, and the chemical operations of digestion and nutrition are maintained; the heat is radiated or conducted from the body into the atmosphere, and again raises the temperature and goes to do other work. At last, so far as our earth is concerned, it escapes into the stellar spaces, and goes to bless other worlds. In all these operations no heat-force is frittered away and wasted and lost. This is one of the accepted doctrines of physical science. Heat is used bountifully, but economically and without waste.

"Even the inequalities and variations of temperature must be counted economy in the use of heat. The heat of midday is not needed at all hours, and therefore it is not always provided; the heat of summer is not always useful, and is therefore not given; a higher temperature for a part of the year and a part of the day is necessary, and is bestowed. The smallest amount of heat is so disposed as to accomplish the largest result. Keep in mind, then, the economical aspect of God's management of heat.

"I would also have you remember how few are the principles involved in all the ways and means for transporting heat and equalizing temperature. All the various phenomena which we have examined can be brought under two general principles. The first principle or method is the heating and cooling of bodies. Bodies absorb heat; they part with their heat by conduction or radiation. If they are heated and cooled without change of place, heat is transported in time, but not in place. If the body be removed from one place to another between the heating and the cooling or between the cooling and the heating, heat is transported in both time and space. This applies alike to solids, liquids, and gases; each one is a carrier of heat in proportion to its specific heat.

"The second principle or method is the transportation of heat by the change of sensible to latent heat and its restoration to a sensible state. Under this principle there are four cases:

"1. Heat is employed in the evaporation of liquids, and is restored again to use as affecting temperature by the condensation of the vapor.

"2. Heat is employed in liquifying solids, and becomes latent thereby, and returns to the sensible state when the liquid solidifies. These two principles find their grandest application in the changes of water: of this application I have chiefly spoken; but they apply also to other bodies—to metals as well as to liquids.

"3. Heat is rendered latent in the expansion of gases from removal of pressure, and latent heat becomes sensible by the compression of gases.

"4. Heat is employed in the deoxidation of carbonic acid or other combinations of oxygen, and is evolved in combustion. While in the latent condition, heat may be kept without loss for an unlimited period of time or transported from equator to pole. By the various applications of these two general principles, all the different methods of equalizing temperature are determined.

"I would have you remember also that these processes for transporting heat and modifying temperature are not confined to the regular changes of days and seasons and the permanent differences of zones, but apply to every possible difference of temperature. One minute the sun shines out in full splendor; the next, a cloud hides his face and cuts off his fervent beams; the methods employed to soften the heat of the one minute and the chill of the next are the same which equalize the temperature of the seasons. Evaporation carries off the heat from the seething tropics, evaporation carries off the excess of heat from the bodies of animals and men. The same methods are equally efficient upon the grandest and upon the smallest scale.

"In this connection let me give one or two illustrations of the delicacy with which general principles adapt themselves to the minutest circumstances. When the earth is wet, it is fitting that evaporation should go on rapidly and remove the excess of water, but when the ground is drier, it is fitting that evaporation should be checked and the remaining moisture spared. This result is secured not merely by the lack of moisture at the surface, but also by the decreased capacity of the earth for absorbing heat. A dark color absorbs heat more readily than a lighter color, and the earth becomes, as a general rule, darker when wet; and lighter when dry. Moist earth, therefore, receives heat more readily than dry earth, and the excessive moisture is the more rapidly carried off by evaporation.

"Another more interesting illustration is presented by the odor of flowers. In its place I told you that watery vapor hinders the radiation of heat from the earth. Dark heat is absorbed by it. The same is true of other gases, and also of the odors of fragrant substances. A bed of flowers fills the air around with odors. By these odors much of the heat radiated by the earth is stopped. By this means the air around the blooming flowers is warmed. The invisible fragrance raises the temperature and secures for the blooming plants a more genial atmosphere. The Lord provides for the flowers when most of all they need to be cherished by a congenial warmth.

"This completes what I have to say to you upon the subject of heat. I might have gone far more into particulars, and extended these lessons over six months instead of three. We started with the design of finding out whether the works of Nature have anything to say about a wise and good Creator. We could not examine the whole circle of God's works, and therefore chose a single department—that of heat. I will leave yourselves to decide whether we have found marks of divine wisdom and goodness, whether Nature has had anything to say to *us* about a Creator."

"It seems to me," said Samuel, "that if the works of Nature do not show God's goodness and wisdom, it would be hard to tell what works would show them. I think I shall always, after this, look upon the earth and sky with more interest than I have ever felt in them before; I shall always look upon them as having something to do with God."

"We certainly ought," said Mr. Wilton, "to study Nature in such a manner and with such a spirit that we shall be led to reverence and worship the Creator. Some very good men are afraid of scientific study, as if there were something in it to draw men from belief in the Scriptures and the Jehovah revealed in them; and it cannot be denied that not a few unbelievers have tried to find a foundation and a defence for their infidelity in scientific studies; but such men are not made skeptics by earnest and reverent study of God's works: they were unbelievers before and aside from physical studies, and they only try to glorify their rejection of the Bible and Christ by deifying science and the creation and holding them up in opposition to inspired revelations. If ever you find the works of God separating you from God, you may know at once that you misunderstand those works or come to them with a wrong spirit. 'The undevout astronomer,' it has been said, 'is mad,' and the same might, with good reason, be said of every undevout student of physical science.

"In selecting heat for our examination, I did not take the only rich department of Nature's works. The practical chemist would find a richer and broader field of research, and so would the anatomist and animal physiologist, the geologist, or the physical geographer. I purposely chose a comparatively narrow field, in order that our course of study might not become wearisome by its length. You will find ample scope in the fields of natural science for your largest powers, and enough to carry your thoughts reverently to the great Creator and Governor.

"In one respect the study of Nature resembles the study of the Sacred Scriptures. It is a revelation; it is an embodiment of God's thoughts; in it God has expressed himself; and Nature, by most suggestive symbols and types, teaches much more moral truth and spiritual sentiment than some men think. In the brute creation it gives us, in pantomime, all the virtues and graces and all repulsive vices and cruel passions. To this book of Nature we ought to come without prejudice, reverently inquiring what is written therein. We must study it thoroughly and interpret it as we interpret the written word, comparing Scripture with Scripture. It is a great attainment to be able to read and understand the thoughts of God embodied in his works.

"In another respect, the book of Nature and the Sacred Scriptures have very little in common. The Bible is occupied pre-eminently with moral duties and spiritual relationship. Its great themes are sin and salvation. Christ is the great central truth. One might compare the Scriptures to a picture in which one central figure seizes every eye, and by whose radiance the whole picture is filled with light, and that central figure is Christ; or we might compare the Bible to a sublime oratorio, the glorious symphony of the ages; through it all is heard one strain, sweetly exultant as angel voices, faintly heard at first amid the sadness of the fall, but rising still above the terrific bass of Sinai and its ever-repeating echoes, growing more clear and strong upon the harps of the prophets, till its rapturous beauty pours itself triumphant along the plains of Bethlehem. In this revelation of salvation from the guilt and ruin of sin the Bible stands alone. Upon this subject Nature is silent. Salvation by Christ is the gem enshrined in the Scriptures. But what is the setting for this gem? The works of God on the earth and in the heavens. The prophets were men in sympathy with Nature. How David sung the praises of the divine handiwork!—'O Lord, how manifold are thy works; in wisdom hast thou made them all.' 'The heavens declare the glory of

God and the firmament showeth his handiwork. Day unto day uttereth speech, and night unto night showeth knowledge. There is no speech nor language where their voice is not heard.' How Christ unfolded the deepest spiritual truths by the symbols of Nature! But if the casket be so worthy, what shall be said of the gem which is enshrined within? That is the pearl of great price. To that book which speaks in no doubtful voice of deliverance from sin let us turn with increasing reverence; and above all, let us come to him who came to reveal our God, who came to be as well as to make a revelation of God, being himself 'the brightness of his glory and the express image of his person.' I am glad that you all now feel that you know him whom to know is everlasting life."

From these words of Mr. Wilton you will conclude that Ansel has at length found rest in Christ. In another brief chapter I will tell you of his experience, and then bid you adieu.

CHAPTER XVI.

A DAY OF JOY AND GLADNESS.

THE reader has already learned that after Ansel had confessed himself an anxious inquirer and professed himself willing to obey Christ, he remained three or four weeks still in darkness. Others found peace in believing, but he felt no joyful confidence that Christ had received him and forgiven his sins. He sometimes felt almost discouraged, and sometimes was tempted to complain of God for not treating him as favorably as others, or to feel chagrined because others were rejoicing, while he found no light. But he fought against these evil thoughts and insinuations of Satan, and did not flag in his private devotions or cease to confess himself, always and everywhere, an anxious inquirer, still in darkness, but desiring to find the grace of God. If ever he was tempted to push away all concern about salvation and return by force to his former careless state, the words of Christ would come to his mind: "Will ye also go away?" and Peter's answer, "Lord, to whom shall we go? for thou hast the words of eternal life." The alternative, salvation by Christ or the loss of his soul, stared him in the face.

> "I can but perish if I go;
> I am resolved to try;
> For if I stay away, I know,
> I must for ever die."

Great interest was felt for him and much prayer was offered in his behalf, but he seemed to make no progress toward a better state. Mr. Wilton had talked with him, but had failed to discover what it was that hindered his humble acceptance of the grace of Christ. After long and anxious musing upon Ansel's character and surroundings and the previous conversations which he had had with him, Mr. Wilton determined to probe him more fully. For this reason he invited Ansel to his study, where the following conversation transpired:

"Good-morning, my young friend; how do you find yourself to-day?"

"I am feeling, I think, very much as when I was here a week ago."

"Are you becoming discouraged and almost ready to give up all effort to follow Christ?"

"I do sometimes feel very much discouraged, but I am not ready to give up my interest in religion."

"Have you no more enjoyment in reading the Scriptures and in your prayer in secret than you had a week ago?"

"I think that I am trying to do right in doing these things, and I enjoy them better than I should if I felt that I was doing something wrong, but I do not feel as I think a Christian ought to feel."

"Are your thoughts and feelings and opinions about Christ and salvation the same as they were six weeks ago?"

"I think they are very different."

"I am glad to hear that; but can you tell how they are different?"

"At that time I felt that I was a sinner, but was fighting against that feeling. I wished that Christ would let me alone, and that the Holy Spirit would not trouble me. But now I very much wish that I may feel my sins, and that

Christ may come to me and save me. I wish to follow the Spirit."

"Did you expect a month ago that at this time you would be feeling and acting as you now feel and act?"

"No, sir; I meant then to fight it through, and not let anybody know how I felt."

"Do you wish now that you had fought it through, as you proposed, and kept all your feelings to yourself?"

"I am very thankful that I did not keep on hiding my feelings. I almost tremble to think what the result would have been."

"You have said that you wish to spend your life in serving Christ. Does it seem to you a hard and painful work—a work that you would get rid of if you could—or does working for Christ and confessing Christ before men seem attractive?"

"I think his service seems pleasant; there is no other life that seems half as plesant."

"Do you believe that Christ is able to save you?"

"I suppose he is. If he cannot save me, there is no hope for me, for I cannot save myself."

"Do you believe that he is willing to save you?"

"I think he is, if I come to him and trust in him. I suppose he is willing to save all who come to him."

"Are you unwilling to come to him—to trust him and submit to him?"

"I don't know; I have tried to come to Christ, but I have met with no such change as I have always supposed that a Christian ought to have."

"What do you think it is that hinders your coming into light and joy as others have done?"

"I cannot tell. I suppose it must be something or other in myself, but I cannot guess what it is."

"I would like to ask you a few questions which you may think rather close and personal, and which you may find it hard to answer frankly. You know the spiritual adviser, as well as the physician, must first of all find out the condition of the patient."

"I am willing to have you ask any questions you please, and I will try to answer them as well as I can."

"Did you ever think, Ansel, that you were very ambitious?"

"I knew that, like many others, I was a little ambitious, but I never thought that I was very much so."

"Perhaps you were more ambitious than you thought. You know that you would work day and night rather than not stand at the head of every class you were in. On the play-ground you asserted your position as leader in every game. Did you not carry the same idea of being chief into your plans and expectations for the future? You were ambitious of standing the very first whatever course of life you might follow. Was not this so?"

"I don't know: I can't deny it; I think it was."

"It is possible, Ansel, that you are trying to carry the same ambition into the kingdom of Christ. Perhaps you have wished in conversion some brilliant experience which would draw attention to you. Tell me how this is. Would you be satisfied to have a commonplace experience, such as thousands of others have, which would attract no special notice? Have you not formed an idea of the great and brilliant change you must pass through, and are you not refusing to take anything else from the Lord's hands?"

Tears gathered in Ansel's eyes, and his face worked painfully. At length he answered: "Your question is a hard one to answer, but I cannot deny it; I am afraid it is so. I have heard persons tell of the great load of sin like a pack on their shoulders, and of the earth seeming as if it would open and swallow them up, of sleepless nights and unspeakable anguish, and then of light and joy, so that they could never doubt that they were converted. I have been expecting that I was to have such an experience, but I have not seen it. Is it wrong to wish for such an experience?"

"It is certainly wrong to *insist* upon such an experience. God leads each one to himself in his own chosen way. There was but one Saul, whom Christ met and blinded with the dazzling light. As a general rule, when a sinner makes up his mind in what way he will be converted, the Lord will disappoint him. If he fixes in his mind that he will not come to an anxious-seat, or will not confess his feelings till he can say that his sins are pardoned, or will not do anything else, the Lord will very likely bring him to do the very thing he resolved that he would not do. If he attempts to bring his ambitious aspirations into Christ's kingdom, he will be disappointed. 'The first shall be last and the last first.' Men become great in Christian service by counting themselves the least of all, and humbling themselves to become the servants of all. You need to examine yourself in this matter. If you have looked for something great and startling, be contented with something small and commonplace. It is an unspeakable privilege to be brought into Christ's kingdom in any manner. It is sometimes a great blessing to have a very unmarked and plain style of conversion. Such a convert is compelled to look to the truly scriptural evidences of a change of heart instead of resting upon the evidence, often deceptive, of a great and sudden illumination or a fancied voice from heaven. Some of the greatest and best of men have been unable to tell at all the time of their conversion. Richard Baxter could not tell even the year of his change. The best experiences I have known have been those where the converts could tell very little about themselves; they had been doing something else besides looking into themselves to watch the motions of their own thoughts."

"I will try to do as you say. But what kind of evidence am I to look for?"

"The same kind of evidence which you now look for in me or any other Christian. It is not one thing to come to Christ and another thing to follow Christ. The best evidence that a sinner has come to Christ is that he actually follows Christ and serves him. 'By their fruits ye shall know them.' 'Bring forth fruits meet for repentance,' said John the Baptist. Bring forth fruits that show that your thoughts about sin, and about Christ, and about the service of Christ have been changed. Look for the same kind of evidence in yourself that you would look for in any stranger whom you should meet. But above all things take the words of Jesus as true and rest on them; consecrate yourself to Jesus with all the heart; with lowliness of mind hold yourself ready for any work or any sacrifice; you will find that evidences will take care of themselves. When men come into sympathy with Christ, when they believe his words, walk with him, and talk with him, and bear the cross with him, when they enter into a partnership of service and suffering with Christ,—the Spirit bears witness with their spirits that they are born of God."

"I will try to follow your advice, and am very thankful that you have spoken about my ambitious spirit."

"Another caution I wish to give you. Do not think that you, by any methods or by cherishing any spirit, are to make yourself fit to be saved. If you are saved at all, Christ must take you as a sinner, and a great sinner. If you get rid of your spirit of pride, it will be by Christ's saving you from it. Let me also suggest to you that which a consideration of your associations suggested to me, that you may have stumbled at the idea of baptism. You must have heard baptism spoken of very disrespectfully, and it is possible that you may have learned to look upon it as a humiliation and a reproach. You may have recoiled from the thought of submitting to it."

"That was my feeling once, but since I have been willing to have my feelings known I have ceased to be afraid of what those who despise religion may say."

"Be careful now, since you feel that your

sympathies are with the Christian band, that your love of greatness does not lead you to resist the Spirit. Be willing to be small. Be thankful for small gifts. I trust that your present feelings will before long give place to a humble trust, a childlike confidence, and a holy boldness in Christ, and that your usefulness in the kingdom of God will be all the greater because he now requires you in the beginning to trample under foot your budding pride and die to all human ambitions."

When Ansel gave up the idea of a wonderful conversion, a sudden illumination which should bring with it something of éclat, he found that he could understand the Scriptures better and have more enjoyment in his religious duties. While he humbled himself, hoping for little, he found his soul soon filled with a deep, quiet joy.

The next Saturday afternoon was the regular time for the covenant-meeting, and also, according to custom, for hearing the experiences of any who wished to unite with the church by baptism. Ansel, Peter, and Mr. Hume came, along with others, to present themselves to the church. In regard to Mr. Hume there had been much speculation among his former comrades as to what course he would take. Some said: "Mr. Hume will never wet the sole of his foot in that river. Don't you remember how he used to laugh at the idea of being plunged in the river in honor of a dead man? He may talk in meeting, but it is a very different thing to go down into the river with the whole hillside covered with people." Others said: "We can't tell what has come over him, but he will not go back now. He has gone too far to retreat."

Some even ventured to approach Mr. Hume himself with their raillery:

"What do you think now of being dipped in the river in honor of a dead man?"

"I think that I would be willing to be baptized a thousand times if I could recall by that means what I have spoken against baptism."

"And what, Mr. Hume, about the ice water?"

"You know and I know," he answered, "that we always respected those who did not shrink from cold water for Christ's sake. What effeminacy, what more than effeminacy, for a resolute man to hesitate and tremble at baptism! We should be ashamed of such weakness in any worldly matter. I have given you occasion for all your raillery, but as I once was a leader in evil, so I wish that I might lead you to better things."

Ansel, Peter, and the rest gave an account of their religious experiences, and last of all Mr. Hume.

"What leads you," asked Mr. Wilton, "to present yourself to the church, asking for baptism?"

"I think that the love of Christ leads me. I have done a great deal against Christ, and now I wish, if possible, to do something to show my love for him. I come to obey the word and example of Christ by being buried with him in baptism."

They were received for baptism, and the time of administration fixed at half-past twelve o'clock the next day.

The Lord's Day was cold and blustering. Many were disappointed, for they hoped that the day would prove warm and sunny. But the blustering day did not prevent the gathering of a great company by the riverside. As the congregations left the churches they turned their steps toward the place of baptism. Ungodly men turned out, and those who never came to hear the preaching of the gospel flocked together to see the gospel preached by this symbolic service. The word had gone out that Mr. Hume was to be baptized, and this drew together his former associates. At the place chosen the river swept around in a gentle curve and the bank rose up like a magnificent amphitheatre; while just above, the land put out into the water and threw the current upon the opposite side. Here gathered almost the entire population of the village to witness that simple and solemn service which from the days of John the Baptist has thrilled so many hearts. The candidates came warmly clad, brought from their own homes in a close carriage. Gathered there, the little band of Christians, surrounded by so great a cloud of witnesses, first sang the hymn commencing:

> "Thou hast said, exalted Jesus,
> Take thy cross and follow me;
> Shall the word with terror seize us?
> Shall we from the burden flee?
> Lord, I'll take it,
> And rejoicing follow thee."

Then Mr. Wilton read with a voice that reached all the company a few passages from the New Testament which authorized and commanded that service. After that he prayed that the joyful presence of Christ might attend those about to follow him in baptism, that believers might be encouraged, and careless sinners awakened. One by one the converts were buried with Christ, and one by one they came up out of the water, forgetting all else in the joy of obedience. They sang the words consecrated by use at so many riversides:

> "Oh how happy are they
> Who their Saviour obey,
> And have laid up their treasure above!
> Tongue can never express
> The sweet comfort and peace
> Of a soul in its earliest love."

These words found a response in many hearts.

High up upon the river bank were gathered a little knot of mocking unbelievers. One among them, seven years before, had publicly professed his faith in Christ. For a little time he seemed to be treading in the Lord's ways, but falling among evil associates, he not only neglected Christian duties, but became a professed unbeliever. He read infidel books and loaned them to others. He sought to sow the seeds of unbelief wherever he went. Upon this Lord's Day he stood with others profanely mocking at the sacred service. With shivering, tremulous accents he exclaimed, "Poor Harry Gill is very cold; I would not go into the water to please any Christ for five hundred dollars." That young man went home with deep conviction of sin upon him. Two days after, Mr. Wilton was called at ten o'clock at night to visit him. He was trembling like an aspen leaf with his deep anguish of conscience, and for two days and nights his body shook under his fear. Then little by little faith took the place of fear, and hope smiled upon him. He was the next person whom Mr. Wilton baptized.

Look in upon the Christian band assembled that Lord's Day evening. Upon the faces of those who had been baptized there was no sign that the service of that day had been painful; if they had done the duty as a cross, the cross must have been quickly followed by a crown of joy, for every face was radiant with light. Among them was one little girl twelve years of age whose face, as she rose from the water, shone like the face of an angel, and the transfiguration of that moment had hardly begun to fade away. Ansel was peacefully happy, and from the face of Mr. Hume the old look of dissatisfaction was all gone; his soul had entered into rest, and he felt at home. Every one of them testified that it had been the happiest day of his life. They declared themselves willing for Christ's sake to be baptized a hundred times if he commanded. They had already found that "in keeping his commandments there is great reward."

I should be glad, kind reader, to trace with you the Christian course of these disciples through the years that follow. But we must leave them. I am sure, however, that their course will be upward. Their experience was not the mere effervescence of fickle feeling. The word of God germinated in their hearts; they had root in themselves. They believed, they believed the truths of the gospel, and therefore they felt, and therefore they acted. "Whatsoever is born of God overcometh the world," and believing that they were truly born of the Spirit, we are confident that "he which hath begun a good work in them will perform it until the day of Jesus Christ."

www.ingramcontent.com/pod-product-compliance
Lightning Source LLC
Chambersburg PA
CBHW031851220426
43663CB00006B/580